MÖBIUS INVERSION IN PHYSICS

Tsinghua Report and Review in Physics

Series Editor: Bangfen Zhu *(Tsinghua University, China)*

Vol. 1 Möbius Inversion in Physics
by Nanxian Chen

MÖBIUS INVERSION IN PHYSICS

Chen Nanxian
Tsinghua University, China

Published by

World Scientific Publishing Co. Pte. Ltd.
5 Toh Tuck Link, Singapore 596224
USA office: 27 Warren Street, Suite 401-402, Hackensack, NJ 07601
UK office: 57 Shelton Street, Covent Garden, London WC2H 9HE

British Library Cataloguing-in-Publication Data
A catalogue record for this book is available from the British Library.

MÖBIUS INVERSION IN PHYSICS
Tsinghua Report and Review in Physics — Vol. 1

Copyright © 2010 by World Scientific Publishing Co. Pte. Ltd.

All rights reserved. This book, or parts thereof, may not be reproduced in any form or by any means, electronic or mechanical, including photocopying, recording or any information storage and retrieval system now known or to be invented, without written permission from the Publisher.

For photocopying of material in this volume, please pay a copying fee through the Copyright Clearance Center, Inc., 222 Rosewood Drive, Danvers, MA 01923, USA. In this case permission to photocopy is not required from the publisher.

ISBN-13 978-981-4291-62-0
ISBN-10 981-4291-62-5

Printed in Singapore by B & Jo Enterprise Pte Ltd

To my parents Chen Xuan-shan and Lin De-yin.
To my wife He Tong and my daughter Jasmine.

Preface

August Ferdinand Möbius gave three profound gifts to the scientific world: the Möbius strip in topology, the Möbius transformation in projective geometry, and the Möbius inverse formula in number theory. This book focuses on the last, and perhaps the least well known of these gifts, introducing the applications of the Möbius inverse formula to a variety of fields in physics, including mechanics, optics, statistical physics, solid state physics, astrophysics, parallel Fourier transform, and others. In order to give beginners an impression of the range and scope of these applications, several are introduced here at the outset.

(1) It is well known that the Fourier transform expands a square wave by a series of sine/cosine waves as

$$S(t) = \sin t - \frac{1}{3}\sin\frac{t}{3} + \frac{1}{5}\sin\frac{t}{5} - \frac{1}{7}\sin\frac{t}{7} + \frac{1}{9}\sin\frac{t}{9}$$
$$- \frac{1}{11}\sin\frac{t}{11} + \frac{1}{13}\sin\frac{t}{13} - \frac{1}{15}\sin\frac{t}{15} + \ldots$$

The Möbius inversion further shows how to expand a sine/cosine wave in terms of square waves. This is useful for designing non-orthogonal modulations in communication, parallel calculations and signal processing.

(2) Also familiar is that the inverse heat capacity problem aims to extract the vibrational frequency spectrum $g(\nu)$ inside a solid based on the measured $C_v(T)$, the temperature dependence of heat capacity. This can be expressed as an integral equation as

$$C_v(T) = rk \int_0^\infty \frac{(h\nu/kT)^2 \exp h\nu/kT}{(\exp h\nu/kT - 1)^2} g(\nu) d\nu$$

Einstein and Debye had contributed two different solutions for this important problem. The Möbius inversion provides an unexpectedly general solution, which not only unifies Einstein's and Debye's solutions, but also significantly improves upon them. Similarly, and of significance to remote sensing and astrophysics, the Möbius inversion method also solves the inverse blackbody radiation problem.

(3) When the interatomic potentials $\Phi(x)$ are known for a crystal, its cohesive energy under varying pressure and temperature can be evaluated directly by summation. But a more practical problem is how to pick up the interatomic potentials based on *ab initio* data of the cohesive energy curve. Möbius inversion solves this inverse lattice problem, and gives a concise and general solution for pairwise potentials. In principle, the Möbius inversion method is also useful for extracting many body interactions in clusters and solids. A similar procedure is also adaptable to deducing the interatomic potentials of an interfacial system.

In spite of the popular view that the fields are too abstruse to be useful, number theory and discrete mathematics have taken on increasingly important roles in computing and cryptography. To provide an example from personal experience, browsing through the shelves in book stores these days, I notice a growing number of volumes devoted to the fields of discrete mathematics, modern algebra, combinatorics, and elementary number theory, many of which are specifically marked as reference study materials for professional training in highly practical fields such as computer science and communication. This trend linking classical number theory with modern technical practice is inspiring. Yet very few books have related elementary number theory to physics. It is somewhat puzzling that there are so few books on physics addressing number theory, though quantum theory emerged much earlier. This book emphasizes the importance and power of Möbius inversion, a tiny yet important area in number theory, for the real world. The application of Möbius inversion to a variety of central and difficult inverse problems in physics quickly yields unexpected results with deep ramifications.

Stated more generally, this book aims to build a convenient bridge between elementary number theory and applied physics. As a physicist, I intend for this book to be accessible to readers with a general physics background, and the only other prerequisites are a little calculus, some algebra, and the spark of imagination.

This book may also prove beneficial to undergraduate and graduate students in science and engineering, and as supplementary material to mathematical physics, solid state physics and elementary number theory courses.

Illustrations included in the book contain portraits of scientists from stamps and coins of many countries throughout history. I hope they provide pleasant and fruitful breaks during your reading, perhaps triggering further study and reflection on the long, intricate, and multi-faceted history behind the creation and development of many of these theories and applications[1].

The arrangement of this book is as follows. Chapter 1 presents basic mathematical knowledge, including the Möbius–Cesàro inversion formula and a unifying formula for both conventional Möbius inverse formula and Möbius series inversion formula. Chapter 2 addresses applications of the Möbius inversion formula to the inverse boson system problem, such as inverse heat capacity and inverse blackbody radiation problems. Chapter 3 concerns applications of the additive Möbius inversion formula to inverse problems of Fermi systems, including the Fermi integral equation, the relaxation-time spectrum, and the surface absorption problem. This chapter also explores a new method for Fourier deconvolution, and the additive Möbius-Cesàro inversion formula. Chapter 4 demonstrates applications of the Möbius inversion formula to parallel Fourier transform technique, presenting, in particular, a uniform sampling method for the arithmetic Fourier transform. Chapter 5 shows the Möbius inversion formulas on Gaussian's and Eisenstein's integers, as well as corresponding applications such as to VLSI processing. Chapter 6 addresses applications of the Möbius inversion method to 3D lattice systems, interfacial systems and others. Chapter 6 also discusses issues related to the embedded atom method (EAM), the cluster expansion method (CEM), and the Möbius inversion formula on a partially ordered set (POSET). Then a short epilogue is attached.

The order for reading is flexible. For most non-mathematicians, the basic knowledge in sections 1.2-1.5 is necessary. Beyond that single proviso the chapters may be selected and reviewed in any order the reader, with whatever background, prefers. It should be emphasized that this book is specially designed to be accessible and engaging for non-mathematicians, which includes the author himself.

N.X. Chen

---

[1] Curious readers can find supplementary materials on these scientists through http://www-groups.dcs.st-and.ac.uk/ history/BiogIndex.html.

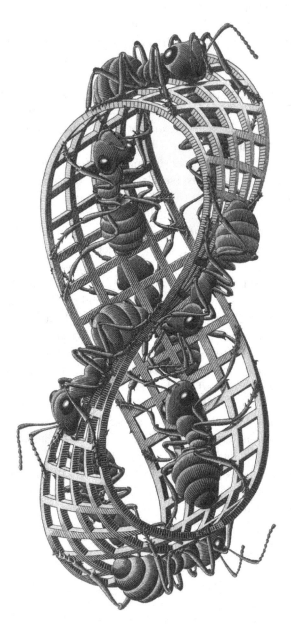

Fig. 0.1 Möbius strip – a combination of aesthetic and scientific truths and contradictions

Acknowledgments

This work was initiated at University of Science and Technology Beijing, and the first edition was finished at Tsinghua University.

Here, the author would like to gratefully acknowledge Menghui Liu, Dunren Guo, Chengbiao Pan, Yuan Wang, and Zhaodou Chen, for their stimulating discussion and persistent help. Without their support, the author would not even have learned of the Möbius function and inversion.

The author also thanks Z.H. Jia, S. Rabii, J.P. Zhang, C.S. Wu, L. Zhang, S.Y. Ren, Z.X. Wang, Gabor Luko, B.F. Zhu, and S.M. Shi for their very kind suggestions. Thanks are due to many others, including E.Q. Rong, T.L. Xie, G.Y. Li, Q. Xie, W.Q. Zhang, S.J. Liu, M. Li, G.B. Ren, Chen Ying, W.X. Cong, J. Shen (Shen Jiang), Y.N. Shen, S. Zhang, Y. Wang, L.Z. Cao, X.L. Wang, Y.C. Wei, Y. Liu, S.J. Hao, Y.M. Kang, X.P. Su, H.Y. Geng, J.Y. Xie, H.Y. Wang, Y. Long, J. Cai, J.G. Guo, W.X. Li, X.Y. Hu, W.X. Su, C. Wang, P. Qian, H.Y. Zhao, G. Liu, Y.D. Wang and Yi Chen for their assistance and sharing in the joys and sorrows of the applications of the Möbius inversion formulas. The author would also like to thank Steven Shi and Zhang Fang at World Scientific Press and Eric Giroux for the combination of enthusiasm and professionalism which they brought to the project. In particular, the author would like to express his deep thanks to the long-term support and encouragement from National Nature Science Foundation in China and National 863/973 Project in China.

Finally, the author thanks in advance all those who send me suggestions, corrections and comments in the future[2].

---

[2] Email address: nanxian@mail.tsinghua.edu.cn

Contents

Preface vii

Acknowledgments xi

List of Figures xix

List of Tables xxiii

1. Basics of Möbius Inversion Formulas   1
   - 1.1 Deriving Möbius Series Inversion Formula with an Example in Physics 1
   - 1.2 Elementary Concepts of Arithmetic Function 8
     - 1.2.1 Definition of an arithmetic function 8
     - 1.2.2 Dirichlet product between arithmetic functions . . 9
     - 1.2.3 All reversible functions as a subset of arithmetic functions 11
   - 1.3 Multiplicative Functions – a Subgroup of the Group $\mathbb{U}$ . . 12
     - 1.3.1 Group $\mathfrak{M}$ of multiplicative functions 12
     - 1.3.2 Unit constant function ι_0 and sum rule of $\mu(n)$ . . 16
     - 1.3.3 Modified Möbius inversion formulas 18
     - 1.3.4 An alternative Möbius series inversion formula . . 21
   - 1.4 Riemann's $\zeta(s)$ and $\mu(n)$ 25
   - 1.5 Möbius, Chebyshev and Modulation Transfer Function . . 28
   - 1.6 Witten Index and Möbius Function 32
   - 1.7 Cesàro–Möbius Inversion Formula 36
   - 1.8 Unification of Eqs. (1.20) and (1.47) 40
   - 1.9 Summary 42
   - 1.10 Supplement – the Seminal Paper of Möbius 42

2. Inverse Problems in Boson Systems — 47

- 2.1 What is an Inverse Problem? — 47
- 2.2 Inverse Blackbody Radiation Problem — 49
  - 2.2.1 Bojarski iteration — 50
  - 2.2.2 The Möbius inversion for the inverse blackbody radiation — 52
- 2.3 Inverse Heat Capacity Problem — 53
  - 2.3.1 Historical background — 53
  - 2.3.2 Montroll solution — 55
  - 2.3.3 The Möbius formula on inverse heat capacity problem — 60
  - 2.3.4 General formula for the low temperature limit — 61
  - 2.3.5 Temperature dependence of Debye frequency — 62
  - 2.3.6 General formula for high temperature limit — 63
  - 2.3.7 Some special relations between $\zeta(s)$ and $\mu(n)$ — 68
- 2.4 Some Inverse Problems Relative to Frequency Spectrum — 73
  - 2.4.1 Inverse spontaneous magnetization problem — 73
  - 2.4.2 Inverse transmissivity problem — 75
- 2.5 Summary — 75

3. Inverse Problems in Fermion Systems — 79

- 3.1 The Arithmetic Functions of the Second Kind — 79
  - 3.1.1 Definition of an arithmetic function of the second kind — 79
  - 3.1.2 Unit function in $\mathbb{A}_2$ — 81
  - 3.1.3 Inverse of an arithmetic function — 81
- 3.2 Möbius Series Inversion Formula of the Second Kind — 82
- 3.3 Möbius Inversion and Fourier Deconvolution — 83
- 3.4 Solution of Fermi Integral Equation — 85
  - 3.4.1 Fermi integral equation — 85
  - 3.4.2 Relaxation-time spectra — 90
  - 3.4.3 Adsorption integral equation with a Langmuir kernel — 91
  - 3.4.4 Generalized Freundlich isotherm — 93
  - 3.4.5 Dubinin–Radushkevich isotherm — 93
  - 3.4.6 Kernel expression by $\delta-$ function — 95
- 3.5 Möbius and Biorthogonality — 96
  - 3.5.1 Chebyshev formulation — 96

		3.5.2	From orthogonality to biorthogonality	99
		3.5.3	Multiplicative dual orthogonality and square wave representation	102
		3.5.4	Multiplicative biorthogonal representation for saw waves	103
	3.6	Construction of Additive Biorthogonality		104
		3.6.1	Basic theorem on additively orthogonal expansion	105
		3.6.2	Derivative biorthogonality from even square waves	107
		3.6.3	Derivative set from triangular wave	112
		3.6.4	Another derivative set by saw wave	114
		3.6.5	Biorthogonal modulation in communication	116
	3.7	Cesàro Inversion Formula of the Second Kind		119
	3.8	Summary		122
4.	Arithmetic Fourier Transform			125
	4.1	Concept of Arithmetic Fourier Transform		125
	4.2	Fundamental Theorem of AFT (Wintner)		127
		4.2.1	Statement of the Wintner theorem	127
		4.2.2	Proof of Eq. (4.2)	127
		4.2.3	Proof of Eq. (4.3)	129
	4.3	The Improvement of Wintner Algorithm by Reed		130
		4.3.1	Two other modified Möbius inverse formulas . . .	130
		4.3.2	Reed's expression	132
	4.4	Fundamental Theorem of AFT (Bruns)		135
		4.4.1	Proof of Eq. (4.41)	136
		4.4.2	The relationship between $a(n), b(n)$ and $B(2n, \alpha)$	137
	4.5	Uniformly Sampling in AFT based on Ramanujan Sum .		140
		4.5.1	What is the Ramanujan sum rule?	140
		4.5.2	Proof of Ramanujan sum rule	141
		4.5.3	Uniformly sampling AFT (USAFT)	142
		4.5.4	Note on application of generalized function	145
	4.6	Summary		146
5.	Inverse Lattice Problems in Low Dimensions			149
	5.1	Concept of Low Dimensional Structures		149
	5.2	Linear Atomic Chains		150
	5.3	Simple Example in a Square Lattice		152
	5.4	Arithmetic Functions on Gaussian Integers		154

		5.4.1	Gaussian integers	154

- 5.4.1 Gaussian integers 154
- 5.4.2 Unit elements, associates with reducible and irreducible integers in $\widetilde{G}$ 154
- 5.4.3 Unique factorization theorem in $\widetilde{G}$ 155
- 5.4.4 Criteria for reducibility 156
- 5.4.5 Procedure for factorization into irreducibles ... 158
- 5.4.6 Sum rule of Möbius functions and Möbius inverse formula 159
- 5.4.7 Coordination numbers in 2D square lattice 160
- 5.4.8 Application to the 2D arithmetic Fourier transform 164
- 5.4.9 Bruns version of 2D AFT and VLSI architecture . 168
- 5.5 2D Hexagonal Lattice and Eisenstein Integers 173
  - 5.5.1 Definition of Eisenstein integers 173
  - 5.5.2 Norm and associates of an Eisenstein integer ... 174
  - 5.5.3 Reducibility of an Eisenstein integer 175
  - 5.5.4 Factorization procedure of an arbitrary Eisenstein integer 176
  - 5.5.5 Möbius inverse formula on Eisenstein integers .. 177
  - 5.5.6 Application to monolayer graphite 179
  - 5.5.7 Coordination number in a hexagonal lattice ... 181
- 5.6 Summary 182

6. **Inverse Lattice Problems**   **183**

- 6.1 A Brief Historical Review 183
- 6.2 3D Lattice Inversion Problem 184
  - 6.2.1 CGE solution 184
  - 6.2.2 Bazant iteration 186
- 6.3 Möbius Inversion for a General 3D Lattice 186
- 6.4 Inversion Formulas for some Common Lattice Structures . 189
  - 6.4.1 Inversion formula for a fcc lattice 189
  - 6.4.2 Inversion formula in a bcc structure 192
  - 6.4.3 Inversion formula for the cross potentials in a $L1_2$ structure 195
- 6.5 Atomistic Analysis of the Field-Ion Microscopy Image of Fe_3Al 197
- 6.6 Interaction between Unlike Atoms in B_1 and B_3 structures 201
  - 6.6.1 Expression based on a cubic crystal cell 201
  - 6.6.2 Expression based on a unit cell 203
- 6.7 The Stability and Phase Transition in NaCl 205

6.8	Inversion of Stretching Curve	209
6.9	Lattice Inversion Technique for Embedded Atom Method	211
6.10	Interatomic Potentials between Atoms across Interface	215
	6.10.1 Interface between two matched rectangular lattices	215
	6.10.2 Metal/MgO interface	217
	6.10.3 Matal/SiC interface	227
6.11	Summary	230

Appendix: Möbius Inverse Formula on a Partially Ordered Set		233
A.1	TOSET	233
A.2	POSET	234
A.3	Interval and Chain	235
A.4	Local Finite POSET	236
A.5	Möbius Function on Locally Finite POSET	237
	A.5.1 Example	238
	A.5.2 Example	239
A.6	Möbius Inverse Formula on Locally Finite POSET	239
	A.6.1 Möbius inverse formula A	239
	A.6.2 Möbius inverse formula B	240
A.7	Principle of Inclusion and Exclusion	241
A.8	Cluster Expansion Method	245

Epilogue	251
Bibliography	253
Index	263

List of Figures

0.1 Möbius strip – a combination of aesthetic and scientific truths and contradictions x

1.1 One dimensional atomic chain. 1
1.2 German mathematician J.W.R. Dedekind (1831–1916) was the last student of Gauss in Göttingen. This is a stamp with an image of Dedekind and unique factorization 4
1.3 August F. Mobius (1790–1864) and Lejeune Dirichlet (1805–1859). 7
1.4 A banknote with the image of Carl Friedrich Gauss (1777–1855). 10
1.5 Leonhard Euler(1707–1783). 15
1.6 Bernhard Riemann(1826–1866) 26
1.7 Pafnuty Chebyshev (1821–1894) 29
1.8 Comparison of sine and bar pattern 29
1.9 Albert Einstein (1879–1955) and Kurt Gödel (1906–1978) . . . 34
1.10 Ernesto Cesàro (1859–1906) 37
1.11 A stamp with a Möbius ring 44

2.1 Jacques Hadamard (1865–1963) 48
2.2 (a)John William Strutt (Lord Rayleigh) (1842–1919), (b) Sir James Hopwood Jeans (1877–1946), (c)Wilhelm Wien (1864–1928) 49
2.3 Image of Max Planck (1858–1947): (a) on a stamp, and (b) in 1901. He said that this radiation law, whenever it is found, will be independent of special bodies and substances and will retain its importance for all times and cultures, even for non-terrestrial and non-human ones. After the great discovery, Planck appeared drained of vitality, exhausted after years of trying to refute his own revolutionary ideas about matter and radiation. 51

2.4	Ludwig Boltzmann (1844–1906) and Satyendra Nath Bose (1894–1974)	53
2.5	Calculated phonon density of states for a sc structure.	54
2.6	(a) A. Einstein (1879–1955) and Mileva (1875–1948) in 1905, (b) Peter Joseph William Debye (1884–1966), a Dutch physicist and physical chemist, (c)Results from Einstein(1907) and Debye(1912).	56
2.7	Elliott Montroll (1916–1983) and G.H. Weiss (1928–)	57
2.8	Pierre-Simon Laplace (1749–1827)	61
2.9	Temperature dependence of Debye frequency	63
2.10	Niels Henrik Abel (1802–1829)	65
2.11	Andrei Nikolaevich Tikhonov (1906–93)	67
2.12	The path of integral	70
2.13	Three kinds of typical inverse problems	76
3.1	Fermi–Dirac distribution	85
3.2	(a) Enrico Fermi (1901–1954), (b) Paul Dirac (1902–1984)	86
3.3	Irving Langmuir (1881–1957)	91
3.4	Norbert Wiener (1894–1964): (a) in Tufts College at the age of eleven, and (b) in his lecture.	93
3.5	(a) Eberhard Hopf (1902–1983). (b) Uzi Landman (1944–)	94
3.6	Triangular waves: (a) Even and (b) odd	97
3.7	Square waves with (a) even and (b) odd	98
3.8	Saw wave	98
3.9	A set of even square waves and its reciprocal set	103
3.10	American Mathematician Walsh (1895–1973) and Walsh function	104
3.11	Set of saw waves and its reciprocal set	105
3.12	Derivative set of square waves and the reciprocal set	108
3.13	A.M. Legendre (1752–1833) and F.W. Bessel (1784–1846)	110
3.14	Derivative set of Legendre functions and the reciprocal set	111
3.15	Derivative set of a saw wave and its reciprocal set	113
3.16	Another derivative set from saw wave and its reciprocal set	115
3.17	Principle of multi-channel communication	117
3.18	The first step of modulation	118
3.19	The first step for demodulation	119
3.20	Filtering: the second step of demodulation	120
4.1	Joseph Fourier (1768–1830) and Fourier series.	125
4.2	(a) James W. Cooley (1926–). (b) John W. Tukey (1915–2000)	126

4.3	Aurel Wintner (1903–1958)	130
4.4	D.W.Tufts (1933–) and I.S.Reed (1923–)	131
4.5	Ernst Heinrich Bruns (1848–1919)	135
4.6	A postage stamp of Srinivasa Ramanujan (1887–1920)	140
5.1	Experiment on Au monatomic chain and an image of a Graphene sheet	150
5.2	(a)Cohesion curve of Au chain. (b)Interatomic potentials in Au chain.	151
5.3	(a)Cohesion curve of Al chain. (b)Interatomic potentials in Al chain.	151
5.4	Gaussian quadrant on a postage stamp with details	155
5.5	VLSI architecture for the present 2D AFT	173
5.6	Gotthold Eisenstein (1823–1852)	174
5.7	A monolayer graphite sheet and the Eisenstein lattice	177
6.1	Left: Max von Laue (1879–1960). Right: W.H. Bragg (1862–1942) and W.L.Bragg (1890–1971).	183
6.2	(a) The face-centered structure and (b) body-centered structure	190
6.3	$L1_2$ structure, and interatomic potentials in Ni_3Al.	195
6.4	Phonon DOS in Ni_3Al with foreign atoms: (a) Pure Ni_3Al, (b) with Pd replacing part of Ni, (c) with Ag replacing part of Al, (d) with Ag replacing part of Ni.	198
6.5	Left: DO_3 structure of Fe_3Al. Right: A top view of a DO_3-type Fe_3Al tip.	200
6.6	Left: B_1 divide into two fcc structures. Right: B_3 into two fcc.	202
6.7	Unit cell of fcc structure	203
6.8	Stretch deformation	210
6.9	Simple interface model	216
6.10	Two ideal structures of fcc(100)/MgO(100)	217
6.11	Several interatomic potentials across Metal/MgO(001) interface.	222
6.12	Polar interface Ag(111)/MgO(111)	224
6.13	Non-zero element distribution in the above matrix.	225
6.14	Interatomic potentials across Ag(111)/MgO(111) interface.	226
6.15	Ideal Au(111)/SiC(111) interface models used for the inversion method. (a) C-terminated, (b) Si-terminated. Each model consists of six metal layers, six Si layers and six C layers. Only a few layers near interface are shown here.	227

6.16 (a)Calculated pair potentials. (b)ab initio adhesion comparing to that from potentials. Here $\Phi_{Au\text{-}Au}$ for bulk Au is also presented for use in metal slab. 228
6.17 Atomic configurations of the metal(111)/SiC(111) interface. (a) SiC(111) surface (top view). The symbols A, B, C and D refer to top site, hollow site, hex site and bridge site, respectively. (c), (d), (e) and (f) are top-site, hollow-site, hex-site and bridge-site structures (side view), respectively. 229
6.18 C-terminated AE distribution of metal(111)/SiC(111) interfaces(metal=Au, Al, Ag and Pt). 230

A.1 Gian Carlo Rota (1932–1999) in 1970. 234
A.2 (a) A diagram of intersection between A and B. (b) A diagram of intersections among A, B and C. 242

List of Tables

1.1	Möbius function.	6
1.2	Coefficients of modulation transfer function	32
4.1	Uniqueness of k and q	128
5.1	Distance function $b(n)$ and coordination function $r(n)$ in 2D square lattice	152
5.2	$(m^2 + n^2)(\text{mod } 4)$	157
5.3	Möbius functions $\mu(\alpha) = \mu(m+in)$ on $\widetilde{G}$	160
5.4	Symmetry analysis of N(x+iy)	161
5.5	Distance function and coordination number	162
5.6	Relative root-mean-square error (RMSE) between input Gaussian function and the reconstructed wave form with a different number of sampling points	172
5.7	Comparison of different methods for calculating a K-point 2D AFT	173
5.8	Congruences analysis N(mod 3) of reducibility	176
5.9	Möbius functions on Eisenstein integers	177
5.10	Parameters for fitting total energy of graphite sheet	179
5.11	The calculated C-C potential with Rose cohesive energy	180
5.12	Comparison of calculated elastic constants and experimental values	181
6.1	$n^2(\text{mod } 8)$	187
6.2	$(m^2 + n^2)(\text{mod } 8)$	187
6.3	$(k^2 + m^2 + n^2)(\text{mod } 8)$	187
6.4	Inversion coefficients for a fcc structure	192
6.5	Inversion coefficients for a bcc structure	193

6.6 The inversion coefficients in $L1_2$ structure 197
6.7 The Morse parameters for $\Phi_{\text{Al-Al}}(x), \Phi_{\text{Ni-Ni}}(x)$ and $\Phi_{\text{Al-Ni}}(x)$. 198
6.8 Site preference of foreign atom in Ni_3Al 199
6.9 The inversion coefficients in DO_3 structure 201
6.10 Sublimation energy of different pure metals 202
6.11 Bonding energy of different surface atoms in Fe_3Al 202
6.12 Inversion coefficients in B_1 structure 204
6.13 Cross inversion coefficients in B_3 structure 206
6.14 Cross inversion coefficients in B1 structure accompanied by B3 structure 208
6.15 Fitting parameters for short-range interactions in LiCl, NaCl, KCl, RbCl 209
6.16 The coefficients h(k) and g(k) for a Metal(bcc)/MgO(001) interface 221
6.17 RSL parameters of interatomic potentials across M/MgO(100) interfaces (M=Ag, Al, Au, Cu) 223
6.18 RSL parameters of interatomic potentials across Ag/MgO(100) interfaces 226
6.19 RSL parameters of two-body potentials across M/SiC(111) interfaces (M=Au, Ag, Pt, Al) 227
6.20 Stillinger–Weber parameters of three-body potentials across M/SiC(111) interfaces (M=Au, Ag, Pt, Al) 229

Chapter 1
Basics of Möbius Inversion Formulas

1.1 Deriving Möbius Series Inversion Formula with an Example in Physics

Among the many ways to introduce the Möbius function and the Möbius inversion formula, we start from an example in physics. Suppose there is an infinite chain shown in Fig. 1.1 with mono-atoms positioned at equal intervals of x. Assume that the interatomic potentials $\Phi(d)$ are pairwise, where $d = nx$ is the distance between the two atoms in a pair, then for each atom in this chain the cohesive energy $E(x)$ can be expressed as

$$E(x) = \frac{1}{2}\sum_{n \neq 0} \Phi(|n|x) \qquad (1.1)$$

or

$$E(x) = \sum_{n=1}^{\infty} \Phi(nx) \qquad (1.2)$$

where nx is the distance between the reference atom located at the origin and its n-th neighboring atom. In most textbooks of solid state physics the reader may have problems calculating $E(x)$ with the known $\Phi(x)$. However, a variety of studies demonstrate how to obtain $\Phi(x)$ using given $E(x)$ based

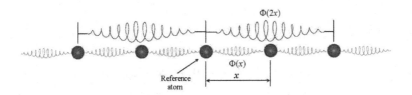

Fig. 1.1 One dimensional atomic chain.

on Eq. (1.2). If we consider the former problem as a direct problem, the latter is inverse with regard to the direct problem.

By using classical Gauss elimination, for example, from (1.2) we have

$$E(x) = \Phi(x) + \Phi(2x) + \Phi(3x) + \Phi(4x) + \cdots$$

and

$$E(2x) = \Phi(2x) + \Phi(4x) + \Phi(6x) + \Phi(8x) + \cdots$$

it follows that

$$E(x) - E(2x) = \Phi(x) + \Phi(3x) + \Phi(5x) + \Phi(7x) + \cdots$$

Next, combining this with

$$E(3x) = \Phi(3x) + \Phi(6x) + \Phi(9x) + \Phi(12x) + \cdots$$

it follows that

$$E(x) - E(2x) - E(3x) = \Phi(x) + \Phi(5x) - \Phi(6x) + \Phi(7x) + \Phi(11x) - \Phi(12x) + \cdots$$

to reason by analogy, finally, the solution of Eq. (1.2) can be expressed as

$$\Phi(x) = E(x) - E(2x) - E(3x) - E(5x) + E(6x) - E(7x) \\ + E(10x) - E(11x) - E(13x) + E(14x) + E(15x) + \cdots$$

If we rewrite it as

$$\Phi(x) = \sum_{n=1}^{\infty} I(n) E(nx) \tag{1.3}$$

then the inversion coefficient $I(n)$ takes 1 when $n = 1, 6, 10, 14, 15, ...$, -1 when $n = 2, 3, 5, 7, 11, \cdots$ and 0 when $n = 4, 8, 9, 12, 16, ...$

Now let us further explore the structure of the inversion coefficients $I(n)$. By using

$$E(x) = \Phi(x) + \Phi(2x) + \Phi(4x) + \cdots$$
$$E(2x) = \Phi(2x) + \Phi(4x) + \Phi(6x) + \cdots$$
$$E(3x) = \Phi(3x) + \Phi(6x) + \Phi(9x) + \cdots$$
$$E(4x) = \Phi(4x) + \Phi(8x) + \Phi(12x) + \cdots$$
$$E(5x) = \Phi(5x) + \Phi(10x) + \Phi(15x) + \cdots$$
$$E(6x) = \Phi(6x) + \Phi(12x) + \Phi(18x) + \cdots$$
$$\cdots$$

then substituting these expressions of $E(x), E(2x), E(3x), ...$ directly into the right hand of Eq. (1.3), one obtains

$$\Phi(x) = I(1)\Phi(x) + \begin{vmatrix} +I(1) \\ +I(2) \end{vmatrix}\Phi(2x) + \begin{vmatrix} +I(1) \\ +I(3) \end{vmatrix}\Phi(3x) + \begin{vmatrix} +I(1) \\ +I(2) \\ +I(4) \end{vmatrix}\Phi(4x)$$

$$+ \begin{vmatrix} +I(1) \\ +I(5) \end{vmatrix}\Phi(5x) + \begin{vmatrix} +I(1) \\ +I(2) \\ +I(3) \\ +I(6) \end{vmatrix}\Phi(6x) + \cdots \quad (1.4)$$

Since the equation just obtained must apply for each value of $\Phi(x)$, it is given that

$$\begin{cases} I(1) = 1 \\ I(1) + I(2) = 0 \\ I(1) + I(3) = 0 \\ I(1) + I(2) + I(4) = 0 \\ I(1) + I(5) = 0 \\ I(1) + I(2) + I(3) + I(6) = 0 \\ \cdots \end{cases} \quad (1.5)$$

through which each $I(n)$ can be extracted by recursive procedure[1].

In order to understand the implications of the above result, we show the solution of Eq. (1.2) in another way. For doing this, let us define an operator T_n such that for any function $g(x)$

$$T_n g(x) = g(nx) \quad \text{with} \quad T_m T_n = T_{mn}. \quad (1.6)$$

Note that the role of T_n is to produce a multiple of the variable of operated function with n. Thus Eq. (1.2) becomes

$$E(x) = \sum_{n=1}^{\infty} T_n \Phi(x) = [\sum_{n=1}^{\infty} T_n]\Phi(x) \equiv \mathbf{T}\Phi(x) \quad (1.7)$$

where $\mathbf{T}$ is defined as a sum of T_n, thus its inverse is given by

$$\Phi(x) = [\sum_{n=1}^{\infty} T_n]^{-1} E(x) = \mathbf{T}^{-1} E(x). \quad (1.8)$$

---

[1] In general, we have $I(1) + I(p) = 0, I(1) + I(p) + I(p^2) = 0, I(1) + I(p) + I(p^2) + \cdots + I(p^k) = 0$, $I(1) + I(p_1) + I(p_2) + I(p_1 p_2) = 0$, and so on. Later we will see that all of these can be attributed as a sum rule $\sum_{d|n} I(d) = \delta_{n,1}$.

Fig. 1.2 German mathematician J.W.R. Dedekind (1831–1916) was the last student of Gauss in Göttingen. This is a stamp with an image of Dedekind and unique factorization

Now the problem is how to simplify the expression of $\mathbf{T}$ or $\mathbf{T}^{-1}$. Recall the fundamental theorem on unique factorization of a positive integer, which states that each natural number n greater than one can be uniquely factorized into primes $p_1, p_2, ..., p_k$ as

$$n = p_1^{\beta_1} p_2^{\beta_2} ... p_s^{\beta_s} \tag{1.9}$$

Similarly, the operator T_n with an arbitrary n can be expressed as a product of T_{p_k}, i.e.,

$$T_n = T_{p_1^{\beta_1} p_2^{\beta_2} ... p_s^{\beta_s}} = T_{p_1^{\beta_1}} T_{p_2^{\beta_2}} ... T_{p_s^{\beta_s}} = (T_{p_1})^{\beta_1} (T_{p_2})^{\beta_2} \cdot (T_{p_s})^{\beta_s} \tag{1.10}$$

in which the term $(T_{p_i})^{\beta_i}$ may be considered as one term picked up from the infinite sum as

$$[1 + T_{p_i} + T_{p_i}^2 + T_{p_i}^3 + T_{p_i}^4 + \cdots].$$

For example, when $n = 90 = 2 \cdot 3^2 \cdot 5$, we have

$$T_{90} = (T_2)(T_3)^2(T_5).$$

The term T_2 can be considered as the second term of the infinite sum

$$[1 + T_2 + T_2^2 + T_2^3 + T_2^4 + \cdots],$$

the term $(T_3)^2$ may be considered as the third term of the infinite sum

$$[1 + T_3 + T_3^2 + T_3^3 + T_3^4 + \cdots],$$

the term T_5 may be considered as the second term of the infinite sum

$$[1 + T_5 + T_5^2 + T_5^3 + T_5^4 + \cdots],$$

and for the parentheses corresponding to other prime, say $p = 7$, we take the first term, $1 = T_7^0$, of the infinite sum

$$[1 + T_7 + T_7^2 + T_7^3 + T_7^4 + \cdots].$$

Now we set a multiplication of infinite number of these parentheses, which runs over all the primes $2, 3, 5, 7, ..., p, ...$ Then all different integers $n's$ can be obtained in the same way. This multiplication is just equal to the total sum of all the operators T_n.

$[1 + T_2 + T_2^2 + T_2^3 + \cdots][1 + T_3 + T_3^2 + T_3^3 + \cdots][1 + T_5 + T_5^2 + T_5^3 + \cdots]$
$[1 + T_7 + T_7^2 + T_7^3 + \cdots] \cdots [1 + T_p + T_p^2 + T_p^3 + \cdots] \cdots$

$$= \sum_{n=1}^{\infty} T_n = \mathbf{T}$$

or

$$\mathbf{T} = \sum_{n=1}^{\infty} T_n = \prod_p \left\{ \sum_{m=0}^{\infty} T_p^m \right\} \tag{1.11}$$

In the above equation, $\mathbf{T}$ becomes a product of infinite parentheses, and each parentheses is a sum as

$$1 + T_p + (T_p)^2 + (T_p)^3 + \cdots = \frac{1}{1 - T_p},$$

and each specific T_n is an product of infinite factors, each factor is only one term from each parentheses. The different selections for taking the term from each parentheses correspond to different n. Now we have

$$\mathbf{T} = \sum_{n=1}^{\infty} T_n = \prod_p \frac{1}{1 - T_p}. \tag{1.12}$$

Hence, the inverse operator $\mathbf{T}^{-1}$ is given as

$$\mathbf{T}^{-1} = \prod_p (1 - T_p). \tag{1.13}$$

Finally, the interatomic potential $\Phi(x)$ in (1.2) can be obtained as

$$\Phi(x) = \mathbf{T}^{-1} E(x) = \prod_p [1 - T_p] E(x)$$
$$= \{(1 - T_2)(1 - T_3)(1 - T_5)(1 - T_7) \cdots \} E(x).$$

In other words,

$$\Phi(x) = E(x) + \sum_{s=1}^{\infty} (-1)^s \sum_{p_1, p_2, ..., p_s} E(p_1 p_2 ... p_s x), \tag{1.14}$$

where the set $p_1, p_2, ..., p_s$ corresponds to a set of s arbitrary distinct primes, and the sum $\sum_{p_1, p_2, ..., p_s}$ runs over all the different sets, each of the sets

Table 1.1 Möbius function.

n	1	2	3	4	5	6	7	8	9	10
$\mu(n)$	1	-1	-1	0	-1	1	-1	0	0	1
n	11	12	13	14	15	16	17	18	19	20
$\mu(n)$	-1	0	-1	1	1	0	-1	0	-1	0
n	21	22	23	24	25	26	27	28	29	30
$\mu(n)$	1	1	-1	0	0	1	0	0	-1	-1

includes several distinct primes. Now let us change the sum notation such that

$$\Phi(x) = E(x) + \sum_{s=1}^{\infty}(-1)^s \sum_{p_1,p_2,\ldots,p_s} E(p_1p_2\ldots p_s x) = \sum_n{}^{*} I(n)E(nx), \quad (1.15)$$

means that n is summed over all integers without repeated factors, for example, n may take $1, 2, 3, 5, 6, 7, 10, \ldots$, not the $8, 9, 12, 16, \ldots$. In other words, n is square-free. Therefore, the inversion coefficient $I(n)$ is given by

$$I(n) = \begin{cases} 1, & n = 1 \\ (-1)^s, & n \text{ is a product of } s \text{ distinct primes.} \end{cases} \quad (1.16)$$

Now let us change the sum notation again such that

$$\sum_n{}^{*} \Rightarrow \sum_{n=1}^{\infty} \quad (1.17)$$

In the left, the sum does not run over those n including repeated factors, and the sum in the right side runs over all the positive integers $\{n\}$. Accordingly, the set of inversion coefficients $\{I(n)\}$ is to be changed to $\{\mu(n)\}$, with

$$\Phi(x) = \sum_{n=1}^{\infty} \mu(n)E(nx) \quad (1.18)$$

where

$$\mu(n) = \begin{cases} 1, & n = 1 \\ (-1)^s, & n = p_1p_2\ldots p_s (p_i \neq p_j, \text{ when } i \neq j) \\ 0, & \text{otherwise.} \end{cases} \quad (1.19)$$

The concrete value of a Möbius function $\mu(n)$ is listed in Table 1.1. Thus, Eq. (1.15) is given explicitly as

$$\begin{aligned}\Phi(x) =& E(x) - E(2x) - E(3x) - E(5x) + E(6x) - E(7x) \\ &+ E(10x) - E(13x) + E(14x) + E(15x) - E(17x) + \cdots\end{aligned}$$

Möbius   Dirichlet

Fig. 1.3   August F. Mobius (1790–1864) and Lejeune Dirichlet (1805–1859).

The jumping of the value of inversion coefficients looks quite interesting. This inversion coefficient $\mu(n)$ is traditionally called the Möbius function in mathematics. Sometimes, Eqs. (1.2) and (1.18) are represented as $E = M\Phi$ and $\Phi = M^{-1}E$ respectively, M is the Möbius series transform, M^{-1} the Möbius series inverse transform[2]. In the present derivation, the Möbius series inversion formula and the Möbius function are introduced simultaneously without any presumption or definition on the Möbius function. Note that in the above deduction the convergence of the infinite sum has been assumed. From the viewpoint of physicists, it is natural to adopt the assumption of convergence since this model represents a reasonable physical case. Note that in practice, it implies some restriction to interatomic potentials. In Chapter 2, we will give an example on how to treat the convergence problem which appears in the inverse heat capacity problem at high temperature. But from the mathematical point of view, it is not appropriate to adopt an ambiguous assumption. Therefore, there is a conventional Möbius inverse formula in most of the textbooks dealing with some special functions in special finite sums which we will discuss later.

---

[2] Möbius was a German astronomer and mathematician. In 1813 Möbius travelled to Göttingen where he studied astronomy under Gauss. Gauss was the director of the Observatory in Göttingen but also, of course, the greatest mathematician of his day, so again Möbius studied under an astronomer whose interests were mathematical. From Göttingen Möbius went to Halle where he studied under Johann Pfaff, Gauss's teacher. Under Pfaff he studied mathematics rather than astronomy, so by this stage Möbius was very firmly working in both fields. Dirichlet was born in Düren, French Empire (now Germany). On Gauss's death in 1855, he was offered his chair at Göttingen, Hanover. [From Full MacTutor biography]

In the above example, we met the unique factorization theorem for any integer, and some functions with integer variables such as $\mu(n), \Phi(nx), E(nx)$. All of these are referred to elementary number theory. Thus, somehow the inverse cohesion problem implies some basic concepts of elementary number theory. Now, let us summarize what we have done in this section with one formula:

Theorem 1.1.
$$G(x) = \sum_{n=1}^{\infty} F(nx) \Rightarrow F(x) = \sum_{n=1}^{\infty} \mu(n) G(nx) \tag{1.20}$$

providing $F(x)$ satisfies
$$\sum_{m,n=1}^{\infty} |F(mnx)| < \infty \tag{1.21a}$$

where $x > 0$. Later in this chapter, we will show the source of the sufficient condition Eq. (1.21a). Note that the inverse theorem of Eq. (1.20) corresponds to another sufficient condition as

$$\sum_{m,n=1}^{\infty} |G(mnx)| < \infty \tag{1.21b}$$

1.2 Elementary Concepts of Arithmetic Function

1.2.1 *Definition of an arithmetic function*

Definition 1.1. An arithmetic function is a function whose domain is $\mathbb{N}$, and whose range is a subset of complex numbers[Pan92; Pan2005; Gio70].

For example, $f(n) = e^{in}$ is an arithmetic function, where $n \in \mathbb{N}, i^2 = -1$. There are many common basic arithmetic functions such as the unit function
$$\Delta(n) = \delta_{n,1}$$
where
$$\delta_{n,k} = \begin{cases} 1, & \text{if } n = k \\ 0, & \text{if } n \neq k \end{cases}$$

and the s-th power function $\iota_s(n) = n^s$, especially the zero-order power function $\iota_0(n) = 1$. This is also called unit constant function. For a quantum system with discrete energy levels, the energy function $\varepsilon(n)$ and corresponding degeneracy function $d(n)$ can also be considered arithmetic functions. In fact, the wave function $\psi_n(x)$ or $\psi(n,x)$ can be considered a continuous function with regard to the variable x, and an arithmetic function with regard to the variable n. This kind of duality exists almost everywhere in mathematical physics. These matters are often ignored and become a blind spot in popular subject. Since 1980 we have entered an "information age", when advanced digital computer and communication technology have greatly boosted the popularity of discrete mathematics, but do physicists in this quantum age apply the tools of discrete mathematics sufficiently?

1.2.2 Dirichlet product between arithmetic functions

In order to determine a set of arithmetic functions it is important to determine the operations between these arithmetic functions.

Definition 1.2. Let $\mathfrak{A}$ denote the set of all arithmetic functions. If f and $g \in \mathfrak{A}$, the Dirichlet product of f and g is a function $f \odot g$ defined by

$$\{f \odot g\}(n) = \sum_{d|n} f(d) g(\frac{n}{d}) \qquad (1.22)$$

where $d|n$ represents that d runs over all positive factors of n including 1 and n. For example, we have

$$\{f \odot g\}(1) = f(1)g(1),$$
$$\{f \odot g\}(2) = f(1)g(2) + f(2)g(1),$$
$$\{f \odot g\}(6) = f(1)g(6) + f(2)g(3) + f(3)g(2) + f(6)g(1).$$

Note that the Dirichlet product is a binary operation on $\mathfrak{A}$. Therefore, $f, g \in \mathfrak{A}$ leads $f \odot g \in \mathfrak{A}$ since the sum and product of complex numbers are still complex numbers. For convenience, sometimes the above definition is written as

$$\{f \odot g\}(n) = \sum_{ab=n} f(a)g(b) \qquad (1.23)$$

Fig. 1.4 A banknote with the image of Carl Friedrich Gauss (1777–1855).

The associativity of the Dirichlet product of arithmetic functions can be shown easily. Therefore, the set of all arithmetic functions is a semigroup[Gio70][3].

Definition 1.3. If a binary operation $\odot$ on an algebraic system $\mathfrak{S}$ satisfies associativity, then the system $\{\mathfrak{S}, \odot\}$ is called a semigroup.

The semigroup of arithmetic functions has other interesting properties, such as commutativity, as follows[4]

---

[3] Note on semigroup: In principle, the semigroup is a more general algebraic system than the group. The main difference between them is that a semigroup does not require the existence of inverse for each operation in it. For example, in a 2D square lattice, if we consider the set of all lattice points, then there are several possible operations that can be defined on this set. For example, representing each lattice point by a vector, then either the scale product or the vector product can not be defined as a binary operation on this set. If representing each lattice point by a complex integer $m + in$, then the set accompanied with common complex product operation becomes a semigroup; meanwhile, the set with common addition operation becomes a group. Also, combining the two operations, the square lattice is called Gaussian integral ring. Different choices correspond to different algebraic structures and reflect different properties of a physical system.

[4] Gauss worked in a wide variety of fields in both mathematics and physics including number theory, analysis, differential geometry, geodesy, magnetism, astronomy and optics. His work has had an immense influence in many areas. He referred to mathematics as "the queen of sciences". Some mathematicians consider the German mathematician Gauss to be the greatest of all time, and almost all consider him to be one of the three greatest, along with Archimedes and Newton; in contrast, he is hardly known to the general public. Within mathematics, number theory was Gauss' first and greatest love; he called it the "Queen of Mathematics" (he has been called the 'Prince of Mathematicians').

1.2.3 All reversible functions as a subset of arithmetic functions

For introducing the reversible function, we have to define the unit function at first.

Definition 1.4. The unit function $\Delta \in \mathfrak{A}$ is defined such that an arbitrary $f \in \mathfrak{A}$ satisfies

$$f \odot \Delta = \Delta \odot f = f \tag{1.24}$$

The existence of unit function can be proven easily[Pan92; Gio70].

Definition 1.5. Assume that $f \in \mathfrak{A}$, if there is a function g such that $f \odot g = g \odot f = \Delta$, then g is called the Dirichlet inverse of f, and denoted as f^{-1}.

The concept of Dirichlet inverse is very useful for many important arithmetic functions.

Theorem 1.2. A necessary and sufficient condition that the Dirichlet inverse of f exists is that $f(1) \neq 0$.

Proof.
(1) Necessity. If f^{-1} exists, then $\{f \odot f^{-1}\}(1) = \Delta(1) = 1$. Thus $f(1) \neq 0$.
(2) Sufficiency. If $f(1) \neq 0$, then we define $g(1) = \frac{1}{f(1)}$. And then we construct $g(2), g(3), \ldots$ one by one by using

$$\sum_{n|k} g(n) f(\frac{k}{n}) = \Delta(k) = \delta_{k,1}$$

This function g agrees with the definition of the Dirichlet inverse of f,

$$f^{-1} \odot f = \sum_{n|k} f^{-1}(n) f(\frac{k}{n}) = \delta_{k,1}$$

Hence, f^{-1} exists if $f(1) \neq 0$. □

Note that if $f(1) = 0$, the above construction procedure for $g(2), g(3), \ldots$ can not be completed. Conventionally, we call both f and f^{-1} reversible functions, and they are mutually dual functions.

Definition 1.6. In the semigroup $\mathfrak{A}$, the subset of all the reversible arithmetic functions forms a group of reversible functions, which is denoted by

$$\mathbb{U} = \{f \in \mathfrak{A} : f(1) \neq 0\} \tag{1.25}$$

It can be proven that the group $\mathbb{U}$ is commutative. Note that the unit function Δ belongs to $\mathbb{U}$.

1.3 Multiplicative Functions – a Subgroup of the Group $\mathbb{U}$

Definition 1.7. An arithmetic function f is called completely multiplicative if $f(mn) = f(m)f(n)$ for all positive integers m and n. An arithmetic function f is called relatively multiplicative if $f(mn) = f(m)f(n)$ whenever m and n are coprime positive integers.

The term "relatively multiplicative" is often simplified as "multiplicative". A completely multiplicative function must be multiplicative, but a multiplicative function is not necessarily completely multiplicative. The multiplicative functions have some important properties: associativity, reversibility, commutativity and divisibility. Now we show the theorem of factorization as follows.

Theorem 1.3. If $f(n)$ is a multiplicative function, then $f(1) = 1$ and

$$n = p_1^{\alpha_1} p_2^{\alpha_2} ... p_k^{\alpha_k} \Rightarrow f(n) = \prod_{i=1}^{k} f(p_i^{\alpha_i}) \tag{1.26}$$

The proof is easy so it is omitted here. Also, denote the set of all multiplicative functions as $\mathfrak{M}$, then $\mathfrak{M} \subset \mathbb{U}$ due to $f(1) = 1 \neq 0$. We will show that this subset $\mathfrak{M}$ also has associativity and reversibility. In other words, it forms a group.

1.3.1 Group $\mathfrak{M}$ of multiplicative functions

Theorem 1.4.

$$\text{If } f, g \in \mathfrak{M}, \text{ then } f \odot g \in \mathfrak{M}. \tag{1.27}$$

Proof. For m and n with $(m, n) = 1$,

$$\{f \odot g\}(mn) = \sum_{d|mn} f(d)g(\frac{mn}{d}) = \sum_{a|m, b|n} f(ab)g(\frac{m}{a}\frac{n}{b})$$

$$= \sum_{a|m, b|n} f(a)f(b)g(\frac{m}{a})g(\frac{n}{b}) = \{\sum_{a|m} f(a)g(\frac{m}{a})\}\{\sum_{b|n} f(b)g(\frac{n}{b})\}$$

$$= \{f \odot g\}(m)\{f \odot g\}(n). \qquad \square$$

Similarly, we can prove the following theorem[Pan92; Gio70].

Theorem 1.5.

If $f \in \mathfrak{M}$, then the Dirichlet inverse f^{-1} of f exists and $f^{-1} \in \mathfrak{M}$. (1.28)

Since $f \in \mathfrak{M} \subset \mathbb{U}$, there must be $f^{-1} \in \mathbb{U}$ and $f^{-1}(1) = 1$. However, $f^{-1} \in \mathbb{U}$ is not equivalent to $f^{-1} \in \mathfrak{M}$. Once $f^{-1} \in \mathfrak{M}$ is proven, one can make sure that the set of all multiplicative functions $\mathfrak{M}$ forms a group. In the following paragraph we see many useful multiplicative arithmetic functions. But, beginners may skip this in the first reading.

Definition 1.8. The s-th power function $\iota_s(n)$ is defined as

$$\iota_s(n) = n^s \qquad (1.29)$$

the power function is obviously multiplicative.

Definition 1.9. Define the divisor function $\tau(n)$ as

$$\tau(n) = \sum_{d|n} 1 \qquad (1.30)$$

which is the number of positive divisors of $n \in \mathbb{N}$.

It is easy to prove that

$$\tau(1) = 1, \text{ and } \tau(p^\beta) = \beta + 1$$

In general,

$$\tau(\prod_i p_i^\beta) = \prod_i (\beta_i + 1) \qquad (1.31)$$

Obviously, $\tau(n)$ is multiplicative.

Definition 1.10. Define the sum function $\sigma(n)$ as

$$\sigma(n) = \sum_{d|n} d \qquad (1.32)$$

which is the sum of positive divisors of $n \in \mathbb{N}$.

It is easy to prove that

$$\sigma(1) = 1, \text{ and } \sigma(p^\beta) = 1 + p + p^2 + \cdots + p^\beta = \frac{p^{\beta+1} - 1}{p - 1}.$$

In general,

$$\sigma(\prod_i p_i^\beta) = \prod_i \frac{p_i^{\beta_i+1} - 1}{p_i - 1} \qquad (1.33)$$

This is also multiplicative.

Now let us show two non-multiplicative examples, and how they are combined into a multiplicative function. Define $\tau_1(n)$ and $\tau_3(n)$ the number of factors of n as $4k+1$ and $4k+3$ respectively, i.e.,

$$\tau_1(n) = \sum_{\substack{d|n \\ d \equiv 1 (\bmod\ 4)}} 1 \qquad (1.34)$$

and

$$\tau_3(n) = \sum_{\substack{d|n \\ d \equiv 3 (\bmod\ 4)}} 1 \qquad (1.35)$$

Obviously, there exists no inverse of $\tau_3(n)$ due to $\tau_3(1) = 0$, and thus the arithmetic function $\tau_3(n)$ is not multiplicative. $\tau_1(1) = 1 \neq 0$, thus there exists inverse of $\tau_1(n)$, but $\tau_1(n)$ is not multiplicative too due to, for instance, $\tau_1(3) = 1, \tau_1(7) = 1, \tau_1(21) = 2 \neq \tau_1(3)\tau_1(7)$.

Considering that

$$\begin{cases} (4k_1+1)(4k_2+1) = 4K_1 + 1 \\ (4k_1+3)(4k_2+3) = 4K_2 + 1 \\ (4k_1+1)(4k_2+3) = 4K_3 + 3 \end{cases} \qquad (1.36)$$

if $(a,b) = 1$, then we have

$$\tau_1(ab) = \tau_1(a)\tau_1(b) + \tau_3(a)\tau_3(b).$$

$\tau_1(ab)$ in the left side of the above expression is the number of divisors d, as $4k+1$, of ab; the two terms in the right side indicates their source as shown in Eq. (1.36). Similarly, it can be given that

$$\tau_3(ab) = \tau_1(a)\tau_3(b) + \tau_3(a)\tau_1(b).$$

Fig. 1.5 Leonhard Euler(1707–1783).

Note that neither $\tau_1(n)$ nor $\tau_3(n)$ is multiplicative. Now a new multiplicative function $\mathfrak{F}(n)$ is introduced as

Definition 1.11.

$$\mathfrak{F}(n) = \tau_1(n) - \tau_3(n) \qquad (1.37)$$

The multiplicative property of $\mathfrak{F}(n)$ can be proven as follows.

Let $(a, b) = 1$, then

$$\mathfrak{F}(ab) = \tau_1(ab) - \tau_3(ab)$$
$$= \tau_1(a)\tau_1(b) + \tau_3(a)\tau_3(b) - \{\tau_1(a)\tau_3(b) + \tau_3(a)\tau_1(b)\}$$
$$= \{\tau_1(a) - \tau_3(a)\}\{\tau_1(b) - \tau_3(b)\} = \mathfrak{F}(a)\mathfrak{F}(b)$$

In other words, $\mathfrak{F}(n) = \tau_1(n) - \tau_3(n)$ is multiplicative, although both $\tau_1(n)$ and $\tau_3(n)$ are not. In Chapter 5 we will meet $\mathfrak{F}(n)$ again.

And the generalized sigma function $\sigma_s(n)$ is defined as

$$\sigma_s(n) = \iota_0 \odot \iota_s \qquad (1.38)$$

$\sigma_s(n)$ is a multiplicative. Also

$$\sigma_s(n) = \{\iota_s \odot \iota_0\}(n) = \sum_{d|n} \iota_s(d)\iota_0(\frac{n}{d}) = \sum_{d|n} d^s. \qquad (1.39)$$

It means that $\sigma_s(n)$ is equal to the sum of d to the $s - th$ power, and d runs over all the positive factors of n. In particular, when $s = 1$, $\sigma_1(n)$ represents the sum of all positive factors of n.

Euler[5] $\varphi-$ function can be defined as the Dirichlet product of Möbius function and natural number(1-st order power function) as

$$\varphi = \iota_1 \odot \mu. \quad (1.40)$$

In order to obtain $\varphi(n)$ one only needs to know $\varphi(p^\beta)$ due to its multiplicative property. Then

$$\varphi(p^\beta) = \sum_{d|p^\beta} \iota_1(d)\mu(\frac{p^\beta}{d}) = \sum_{j=0}^{\beta} \iota_1(p^{\beta-j})\mu(p^j)$$
$$= \iota_1(p^\beta)\mu(1) + \iota_1(p^{\beta-1})\mu(p)$$
$$= p^\beta - p^{\beta-1} = p^\beta(1 - \frac{1}{p}) \quad (1.41)$$

1.3.2 Unit constant function ι_0 and sum rule of $\mu(n)$

Unit constant function is defined as $\iota_0(n) = 1$ for all $n \in \mathbb{N}$. Evidently, ι_0 is a multiplicative arithmetic function. Now let us try to see its Dirichlet inverse ι_0^{-1}. Obviously, $\iota_0^{-1}(n)$ satisfies

$$\sum_{n|k} \iota_0(\frac{k}{n})\iota_0^{-1}(n) = \delta_{k,1} \quad (1.42)$$

or

$$\sum_{n|k} \iota_0^{-1}(n) = \delta_{k,1} \quad (1.43)$$

We are going to list all the values of $\iota_0^{-1}(n)$. Considering that $\iota_0^{-1}(n)$ is also a multiplicative function, we only need to solve $\iota_0^{-1}(p^\beta)$ for obtaining $\iota_0^{-1}(n)$.

(1) $\beta = 1$:

$$0 = \delta_{p,1} = \iota_0^{-1}(1) + \iota_0^{-1}(p) = 1 + \iota_0^{-1}(p) \implies \iota_0^{-1}(p) = -1$$

(2) $\beta = 2$:

$$0 = \delta_{p^2,1} = \iota_0^{-1}(1) + \iota_0^{-1}(p) + \iota_0^{-1}(p^2) \implies \iota_0^{-1}(p^2) = 0$$

---

[5]Leonhard Paul Euler (1707–1783) was a pioneering Swiss mathematician and physicist who spent most of his life in Russia and Germany. Euler made important discoveries in fields as diverse as calculus and graph theory. He also introduced much of the modern mathematical terminology and notation, particularly for mathematical analysis, such as the notion of a mathematical function. He is also renowned for his work in mechanics, fluid dynamics, optics, and astronomy. [From Wikipedia]

By using induction, it is given by

$$\iota_0^{-1}(p^\beta) = \begin{cases} 1, & \beta = 0 \\ -1, & \beta = 1 \\ 0, & \beta > 1 \end{cases} \tag{1.44}$$

Hence, the general expression of $\iota_0^{-1}(n)$ is given by

$$\iota_0^{-1}(n) = \begin{cases} 1, & \text{if } n = 1 \\ (-1)^s, & \text{if } n = p_1 p_2 \dots p_s \\ 0, & \text{if } p^2 | n \end{cases} \tag{1.45}$$

Therefore, the Dirichlet inverse $\iota_0^{-1}(n)$ is just the same as the Möbius function $\mu(n)$ in Eq. (1.19). In other words, the Möbius function μ is the Dirichlet inverse of the unit constant function ι_0, or

$$\mu = \iota_0^{-1} \quad \text{and} \quad \iota_0 = \mu^{-1} \tag{1.46}$$

From this, we have the classical Möbius inverse formula as

Theorem 1.6.

$$g(n) = \sum_{d|n} f(d) \Leftrightarrow f(n) = \sum_{d|n} \mu(d) g(\frac{n}{d}) \tag{1.47}$$

Proof. The left part in Eq. (1.47) can be expressed as

$$g(n) = \sum_{d|n} f(d) \iota_0(\frac{n}{d})$$

or

$$g = f \odot \iota_0$$

From this we have

$$g \odot \mu = f \odot \iota_0 \odot \mu = f$$

That is just the right side of Eq. (1.47). Similarly, one can prove the inverse theorem. □

Obviously, there is no convergent problem involved in the classical Möbius inverse formula Eq. (1.47). Note that Eq. (1.43) is also called the sum rule of Möbius functions, which is conventionally written as

Theorem 1.7.

$$\sum_{d|n} \mu(d) = \delta_{n,1} \tag{1.48}$$

where the sum runs over all the positive divisors of n. The sum rule Eq. (1.48) is essential for Möbius function. Note that a surprising variety of Möbius inversion formulas and Möbius series inversion formulas can be obtained by modifying this sum rule. And the core part of the Möbius inversion formulas is the multiplicity of relative functions (see Theorem 1.4, which exploits the spirit of conventional Möbius inversion.

Now let us show another "proof" of Eq. (1.20) by using the sum rule.

$$\sum_{n=1}^{\infty} \mu(n)G(nx)$$
$$= \sum_{n=1}^{\infty} \mu(n) \sum_{m=1}^{\infty} F(mnx)$$
$$= \sum_{k=1}^{\infty} \left\{ \sum_{n|k} \mu(n) \right\} F(kx)$$
$$= \sum_{k=1}^{\infty} \delta_{k,1} F(kx) = F(x)$$

In this deduction commutativity of terms in the double summations is required, in order words, the double summation has to be unconditionally convergent for the Möbius series inversion formula. Note that Eq. (1.21a) is a sufficient condition for the proof. This might be convenient for most physicists who are familiar with the concept of the absolute convergence[Kno28].

1.3.3 Modified Möbius inversion formulas

Similar to Eq. (1.20), we have

$$f(x) = \sum_{n=1}^{\infty} g(x^n) \Leftrightarrow g(x) = \sum_{n=1}^{\infty} \mu(n) f(x^n). \tag{1.49}$$

For example, assuming that

$$f(x) = \frac{x}{1-x} = x + x^2 + x^3 + x^4 + \cdots \tag{1.50}$$

the corresponding $g(x) = x$, then

$$x = \sum_{n=1}^{\infty} \mu(n) \frac{x^n}{1-x^n} = \frac{x}{1-x} - \frac{x^2}{1-x^2} - \frac{x^3}{1-x^3} - \frac{x^5}{1-x^5} +$$
$$+ \frac{x^6}{1-x^6} - \frac{x^7}{1-x^7} + \frac{x^{10}}{1-x^{10}} - \frac{x^{11}}{1-x^{11}} - \frac{x^{13}}{1-x^{13}} + \cdots \tag{1.51}$$

This interesting example can demonstrate an important relation in statistical physics. It is well known that for a phonon system (or photon system) with vanished chemical potential, the classical Boltzmann[6] distribution and the quantum Bose-Einstein distribution can be written as

$$f_{Bz}(\frac{h\nu}{kT}) = e^{-\frac{h\nu}{kT}} \tag{1.52}$$

and

$$f_{BE}(\frac{h\nu}{kT}) = \frac{1}{e^{\frac{h\nu}{kT}} - 1} = \sum_{n=1}^{\infty} f_{Bz}(n\frac{h\nu}{kT}) \tag{1.53}$$

Therefore,

$$f_{Bz}(\frac{h\nu}{kT}) = \sum_{n=1}^{\infty} \mu(n) f_{BE}(n\frac{h\nu}{kT}) \tag{1.54}$$

Eq. (1.49) can be also modified to

Theorem 1.8.

$$f(x) = \prod_{n=1}^{\infty} g(x^n) \Leftrightarrow g(x) = \prod_{n=1}^{\infty} f(x^n)^{\mu(n)}. \tag{1.55}$$

---

[6]Ludwig Boltzmann(1844–1906) was an Austria physicist. His father was a taxation official. Boltzmann was awarded a doctorate from the University of Vienna in 1866 for a thesis on the kinetic theory of gases. Boltzmann's fame is based on his invention of statistical mechanics. Boltzmann obtained the Maxwell-Boltzmann distribution in 1871, namely the average energy of motion of a molecule is the same for each direction. He was one of the first to recognise the importance of Maxwell's electromagnetic theory. Boltzmann's ideas were not accepted by many scientists. In 1895, at a scientific meeting in Lbeck, Wilhelm Ostwald (1853–1932, a Baltic German chemist) presented a paper in which he stated:- *The actual irreversibility of natural phenomena thus proves the existence of processes that cannot be described by mechanical equations, and with this the verdict on scientific materialism is settled.* Attacks on his work continued and he began to feel that his life's work was about to collapse despite his defence of his theories. Depressed and in bad health, Boltzmann committed suicide just before experiment verified his work. On holiday with his wife and daughter at the Bay of Duino near Trieste, he hanged himself while his wife and daughter were swimming. However the cause of his suicide may have been wrongly attributed to the lack of acceptance of his ideas. We will never know the real cause which may have been the result of mental illness causing his depression. [From Wikipedia]

Proof.

$$\prod_{n=1}^{\infty} f(x^n)^{\mu(n)} = \prod_{n=1}^{\infty} \prod_{m=1}^{\infty} g([x^n]^m)^{\mu(n)}$$

$$= \prod_{k=1}^{\infty} \prod_{n|k} g(x^k)^{\mu(n)} = \prod_{k=1}^{\infty} g(x^k)^{\sum_{n|k} \mu(n)}$$

$$= \prod_{k=1}^{\infty} g(x^k)^{\delta_{k1}} = g(x). \qquad \square$$

Similarly, Eq. (1.20) can be also extended to

Theorem 1.9.

$$f(x) = \sum_{n=1}^{\infty} r(n)g(x^n) \Leftrightarrow g(x) = \sum_{n=1}^{\infty} \mu(n)r(n)f(x^n) \tag{1.56}$$

where $r(n)$ is a completely multiplicative function.

There is an interesting example in the Möbius origin paper: when $|x| < 1$, it follows that

$$f(x) = -\log(1-x) = x + \frac{1}{2}x^2 + \frac{1}{3}x^3 + \cdots \tag{1.57}$$

Taking $r(n) = 1/n$, $g(x^n) = x^n$, it is given that

$$x = -\log(1-x) + \frac{1}{2}\log(1-x^2) + \frac{1}{3}\log(1-x^3)$$

$$+ \frac{1}{5}\log(1-x^5) - \frac{1}{6}\log(1-x^6) + \cdots \tag{1.58}$$

hence,

$$e^x = (1-x)^{-1}(1-x^2)^{1/2}(1-x^3)^{1/3}(1-x^5)^{1/5}(1-x^6)^{-1/6}\cdots \tag{1.59}$$

In fact, this was an example in the seminal paper of Möbius in 1832. Möbius' pioneering idea was to check which function with Taylor's expansion coefficients forming a multiplicative semigroup.

This kind of procedure will be useful for further study. Here, we take two other theorems as follows.

Theorem 1.10.

$$g_n(x) = \sum_{d|n} r(d) f_{\frac{n}{d}}(x) \Leftrightarrow f_n(x) = \sum_{d|n} I(d) g_{\frac{n}{d}}(x) \tag{1.60}$$

where $I(d)$ satisfies

$$\sum_{mn=k} I(n)r(m) = \delta_{k,1} \tag{1.61}$$

Theorem 1.11.

$$G(x) = \sum_{n=1}^{\infty} r(n)F(nx) \Leftrightarrow F(x) = \sum_{n=1}^{\infty} I(n)G(nx) \tag{1.62}$$

where $I = r^{-1}$ is as before. Equation (1.62) is called generalized Möbius inverse formula. It can be shown that one sufficient condition for Eq. (1.62) is $\sum_{m,n=1}^{\infty} |r(m)I(n)F(mnx)| < \infty$ and $\sum_{m,n=1}^{\infty} |r(m)I(n)G(mnx)| < \infty$.

Also, there is a similar formula without the convergent problem as follows.

Theorem 1.12.

$$g(n) = \sum_{d|n} r(d)f(\frac{n}{d}) \Leftrightarrow f(n) = \sum_{d|n} I(d)g(\frac{n}{d}) \tag{1.63}$$

where $I(d)$ satisfies (1.61) too.

From the above we can see that the center of variety of Möbius inversion formulas is the sum rule, and essentially the sum rule is a mutually reversible relation $r \odot r^{-1} = \Delta$. This is also called duality or reciprocal relation.

1.3.4 An alternative Möbius series inversion formula

1.3.4.1 An alternative Möbius series inversion

This can be solved by Eq. (1.63) directly. However, here we introduce another Möbius inversion theorem[Che91]

Theorem 1.13.

$$F(x) = \sum_{n=1}^{\infty} (-1)^{n+1} f(nx) \Leftrightarrow f(x) = \sum_{n=1}^{\infty} \sum_{m=0}^{\infty} 2^m \mu(n) F(2^m nx) \tag{1.64}$$

Proof. Noted that

$$\sum_{k=1}^{\infty}(-1)^{k+1}f(2k) = \sum_{2\nmid k}f(2k) - \sum_{2|k}f(2k)$$

$$= \sum_{k=1}^{\infty}f(2k) - 2\sum_{2|k}f(2k)$$

$$= \sum_{k=1}^{\infty}f(2k) - 2\sum_{k=1}^{\infty}f(2^2 k)$$

and

$$\sum_{m=0}^{\infty}g(2^m) = \sum_{m=1}^{\infty}g(2^{m-1}),$$

then it is given that

$$\sum_{n=1}^{\infty}\sum_{m=0}^{\infty}2^m\mu(n)F(2^m nx) = \sum_{k,m,n=1}^{\infty}(-1)^{k+1}2^{m-1}\mu(n)f(2^{m-1}nkx)$$

$$= \sum_{n=1}^{\infty}\mu(n)[\sum_{k,m=1}^{\infty}2^{m-1}f(2^{m-1}nkx) - \sum_{2|k}\sum_{m=1}^{\infty}2^{m-1}f(2^{2m-1}nkx)]$$

$$= \sum_{n=1}^{\infty}\mu(n)[\sum_{k,m=1}^{\infty}2^{m-1}f(2^{m-1}nkx) - \sum_{k,m=1}^{\infty}2^m f(2^{2m}nkx)]$$

$$= \sum_{n=1}^{\infty}\mu(n)[\sum_{k=1}^{\infty}f(nkx)] = \sum_{k=1}^{\infty}C(kx) = f(x)$$

where $C(x) = \sum\mu(n)f(nx) \Rightarrow f(n) = \sum C(nx)$ is used. $\square$

1.3.4.2 Madelung constant in a linear ionic chain

Suppose that the crystal consists of positive and negative ions lying on a line alternatively with equal spacing corresponding to lattice constant of $2x$. Therefore, the total energy in this case becomes

$$\mathbb{E}(x) = 2\sum_{n=1}^{\infty}(-1)^{n+1}\Phi(nx) \tag{1.65}$$

The question is how to obtain the pair interaction $\Phi(x)$ between ions based on measurable or calculable average potential $E(x)$. From the last subsection, the pairwise interaction is given as

$$\Phi(x) = \frac{1}{2}\sum_{m,n=1}^{\infty}2^{m-1}\mu(n)\mathbb{E}(2^{m-1}nx) \tag{1.66}$$

If the ion–ion interactions are of pure Coulomb type without screening effect, the above result could be used to evaluate the Madelung constant of a linear ionic chain. To do this, first of all, it is assumed that the elementary interaction takes the form of

$$\Phi(x) = 1/x^s$$

and the cohesive energy becomes

$$\mathbb{E}(x) = \frac{M_s}{x^s}$$

where the Madelung constant M_s is to be determined.
Then it is given that

$$\frac{M_s}{2} \sum_{m=1}^{\infty} 2^{(m-1)(1-s)} \sum_{n=1}^{\infty} \mu(n)\frac{1}{n^s} = 1.$$

Note that the sum of a series with equal ratio is

$$\sum_{m=1}^{\infty} 2^{(m-1)(1-s)} = \frac{1}{1 - 2^{1-s}}$$

and

$$\lim_{s \to 1} \sum_{n=1}^{\infty} \frac{\mu(n)}{n^s} = 0 \quad \text{and} \quad \lim_{s \to 1}(1 - 2^{1-s}) = 0.$$

The Madelung constant M is equal to

$$M = \lim_{s \to 1} M_s = 2 \lim_{s \to 1} \frac{1 - 2^{1-s}}{\sum_{n=1}^{\infty} \mu(n)\frac{1}{n^s}} = \frac{0}{0}.$$

Therefore, the L'Hopatal rule is used to get

$$M = 2 \frac{\lim_{s \to 1} \frac{d}{ds}(1 - 2^{1-s})}{\lim_{s \to 1} \frac{d}{ds} \sum_{n=1}^{\infty} \mu(n)\frac{1}{n^s}} = 2\ln 2 \qquad (1.67)$$

since

$$\lim_{s \to 1} \frac{d}{ds}(1 - 2^{1-s}) = \lim_{s \to 1} \frac{d}{ds}\{1 - e^{(1-s)\ln 2}\} = -2^{1-s} \cdot (-\ln 2) = \ln 2$$

and

$$\lim_{s \to 1} \sum_{n=1}^{\infty} \mu(n) \frac{d}{ds} n^{-s} = -\sum_{n=1}^{\infty} \mu(n)\frac{\ln n}{n} = 1.$$

This result is expected by Kittel[Kit76].

1.3.4.3 An inverse problem for intrinsic semiconductors

Now let us apply the above theorem to solve the fermion integral equation for a intrinsic semiconductor

$$n(T) = \int_{E_c}^{\infty} \frac{g(E)dE}{1 + e^{E-E_F/kT}} \tag{1.68}$$

where E_F is the fermi level located at $\frac{1}{2}(E_c + E_v)$, while E_c and E_v represent the bottom of conduction band and the top of valance band respectively. The question is to extract the density of states $g(E)$ from the measurable carrier density $n(T)$.

Introducing the coldness $u = 1/kT$ and defining

$$G(E) = g(E + E_c) \tag{1.69}$$

then

$$n(\frac{1}{ku}) = \int_{E_c}^{\infty} \frac{g(E)dE}{1 + exp\{E - E_F\}u}$$

$$= \int_0^{\infty} dE\, G(E) \sum_{n=1}^{\infty} (-1)^{n+1} \exp\{-nu(E + E_c - E_F)\}$$

$$= \sum_{n=1}^{\infty} (-1)^{n+1} \exp[-nu(E_c - E_F)] \int_0^{\infty} dE\, G(E) \exp(-nuE)$$

or

$$n(\frac{1}{ku}) = \sum_{n=1}^{\infty} (-1)^{n+1} \exp[-nu(E_c - E_F)] \mathcal{L}[G(E); E \to nu] \tag{1.70}$$

Note that in experiment E_F is weakly dependent on temperature T as E_c changes, the reasoning here only requires $E_c - E_F$ being constant. From alternative series inversion it is given that

$$\mathcal{L}[G(E); E \to u] = \exp[u(E_c - EF)] \sum_{m=0}^{\infty} \sum_{n=1}^{\infty} 2^m \mu(n) n(\frac{1}{2^m nu}) \tag{1.71}$$

If $n(T)$ is known, then $G(E)$ or $g(E)$ can be obtained from Eq. (1.71). For example, if the density of n-type carriers can be expressed as

$$n(T) \approx C(kT)^{3/2} e^{-(E_c - E_F)/kT}$$

or

$$n(\frac{1}{kT}) = Cu^{-3/2} e^{-(E_c - E_F)u} \tag{1.72}$$

where C is constant. Then

$$\mathcal{L}[G(E); E \to u] = Cu^{-3/2} + C(2u)^{-3/2}e^{-(E_c-E_F)} - C(3u)^{-3/2}e^{-2(E_c-E_F)} + \cdots$$

By using the inverse Laplace operation, we have

$$G(E) = C_1 E^{1/2} + C_2(E - E_c + E_F)^{1/2} - C_3(E - 2E_c + 2E_F)^{1/2} + \cdots \quad (1.73)$$

Finally it is given by

$$\begin{aligned} g(E) &= C_1(E - E_c)^{1/2} H(E - E_c) \\ &+ C_2(E - E_c + E_F)^{1/2} H(E - 2E_c + E_F) + \cdots \end{aligned} \quad (1.74)$$

where $H(E - E_x)$ is the step function defined as

$$H(E - E_x) = \begin{cases} 1 & E \geq E_x \\ 0 & E < E_x \end{cases} \quad (1.75)$$

The first term in Eq. (1.74) represents the parabolic approximation

$$g(E) = C_1(E - E_c)^{1/2} H(E - E_c) \quad (1.76)$$

as in most textbooks, such as Kittel[Kit76]. In a general case, $n(T)$ might be given as a set of numerical data to be fit. One has to be careful to choose the fitting functions such that each of them are inverse Laplace transformable based on experience[Bel66].

1.4 Riemann's $\zeta(s)$ and $\mu(n)$

Riemann's zeta function[7] is defined as[Sch90]

$$\zeta(s) = \sum_{n=1}^{\infty} \frac{1}{n^s}, \quad s > 1 \quad (1.77)$$

Note that only a real value is taken for s here.

Theorem 1.14.

$$\frac{1}{\zeta(s)} = \sum_{n=1}^{\infty} \frac{\mu(n)}{n^s}, \quad s > 1 \quad (1.78)$$

---

[7]Bernhard Riemann was a German mathematician who made important contributions to analysis and differential geometry, some of them enabling the later development of general relativity. He was a student of Dirichlet. [From Wikipedia]

Fig. 1.6 Bernhard Riemann(1826–1866)

Proof.

$$\zeta(s) \cdot \sum_{n=1}^{\infty} \frac{\mu(n)}{n^s} = \left[\sum_{m=1}^{\infty} \frac{1}{m^s}\right] \cdot \sum_{n=1}^{\infty} \frac{\mu(n)}{n^s}$$

$$= \sum_{m,n=1}^{\infty} \frac{\mu(n)}{(mn)^s} = \sum_{k=1}^{\infty} \left[\sum_{n|k} \mu(n)\right] \frac{1}{k^s} = \sum_{k=1}^{\infty} \delta_{k,1} \cdot \frac{1}{k^s} = 1 \qquad \square$$

Theorem 1.15.

$$\frac{1}{\zeta(s)} = \prod_{p} (1 - \frac{1}{p^s}) \qquad (1.79)$$

where p runs over all the primes.

Proof. Applying the identical equation

$$\sum_{n=1}^{\infty} \frac{1}{n^s} = \prod_{p} (1 + \frac{1}{p^s} + \frac{1}{p^{2s}} + \frac{1}{p^{3s}} + \cdots)$$

then

$$\prod_{p} (1 - \frac{1}{p^s}) = \prod_{p} \frac{1}{1 + \frac{1}{p^s} + \frac{1}{p^{2s}} + \frac{1}{p^{3s}} + \cdots} = \frac{1}{\sum_{n=1}^{\infty} \frac{1}{n^s}} = \frac{1}{\zeta(s)}$$

$\square$

By using Riemann's ζ function, we can discuss the distribution of Möbius function[Dav81]. From Table 1.1 we see that, on one hand, the values of Möbius function $\mu(n)$ appear utterly chaotic with respect to n; on the other hand, the Möbius function is completely deterministic; a unique $\mu(n)$ corresponds to each value of n. Now we are going to consider some characters of the distribution. What are the chances that n has no repeated factor, or $\mu(n) \neq 0$? As long as n is not a multiple of 4 or a multiple of 9 or a multiple of 25, or any other square of a prime, $\mu(n) \neq 0$. Now, the probability that a natural number chosen at random is not a multiple of 4 is 3/4, the probability that a natural number chosen at random is not a multiple of 9 is 8/9, the probability that a natural number chosen at random is not a multiple of 25 is 24/25, and so on. Therefore, the probability that an arbitrary number n does not have repeated prime factors is

$$\prod_p \left(1 - \frac{1}{p^2}\right) = \frac{3}{4} \cdot \frac{8}{9} \cdot \frac{24}{25} \cdot \frac{48}{49} \cdots = \frac{1}{\zeta(2)} = \frac{6}{\pi^2} \quad (1.80)$$

By symmetry, the probability that $\mu(n) = 1$ is $3/\pi^2$, the probability that $\mu(n) = -1$ is also equal to $3/\pi^2$, and the probability that $\mu(n) = 0$ is equal to $(1 - \frac{6}{\pi^2})$. Now let us look at a model on phase transition.

It is well known that the canonical partition function for a classical gas of N non-interacting particles with single particle excitation spectrum ε_n is

$$Q_N(T) = \frac{1}{N!}\left\{\sum_{n=1}^{\infty} \exp(-\beta\varepsilon_n)\right\}^N = \frac{1}{N!}\left\{\sum_{n=1}^{\infty} \exp\left(-\frac{\varepsilon_n}{k_B T}\right)\right\}^N \quad (1.81)$$

Suppose that the spectrum is set as[Nin92]

$$\varepsilon_n = \varepsilon_0 \ln n \quad (1.82)$$

then

$$q(T) = \sum_{n=1}^{\infty} \exp(-\frac{\varepsilon_n}{k_B T}) = \sum_{n=1}^{\infty} \exp(-\frac{\varepsilon_0 \ln n}{k_B T})$$

$$= \sum_{n=1}^{\infty} \exp(\ln n^{-\varepsilon_0/k_B T}) = \sum_{n=1}^{\infty} \frac{1}{n^{\varepsilon_0/k_B T}} = \zeta(\varepsilon_0/k_B T) \quad (1.83)$$

Obviously, $T_0 = \varepsilon_0/k_B$ represents a critical temperature since the partition function $q_N(T)|_{T \geq T_0}$ diverges when T exceeds the critical temperature T_0. However, this kind of critical phenomenon does not necessarily happen even if the temperature T, as a parameter in the statistical distribution, is reduced or being negative. This might be a way to describe an overheating phenomenon or a meta-stable state. Later, we will show a modified relation as

$$\sum_{n=1}^{\infty} \mu(n)n^{2k-1} \longrightarrow \lim_{x\to 1^-} \sum_{n=1}^{\infty} \mu(n)n^{2k-1}x^n = \frac{1}{\zeta(1-2k)} \qquad (1.84)$$

In other words, when $T = -|T|$ being negative even integer, we have

$$\sum_n n^{\varepsilon_0/k_B T} = \zeta(1 - \varepsilon_0/k_B|T|)$$

Note that the ζ function is related to Bernoulli number B_n[Gra80] as

$$2\zeta(2m) = (-1)^{m+1}\frac{(2\pi)^{2m}}{(2m)!}B_{2m}, \quad m = 1, 2, 3, \ldots$$

This is the Euler's result. For example, from

$$B_2 = 1/6, \ B = -1/30 \ \text{and} \ B_6 = 1/42$$

we have

$$\zeta(2) = \pi^2/6, \ \zeta(4) = \pi^4/90, \ \text{and} \ \zeta(6) = \pi^6/945$$

1.5 Möbius, Chebyshev and Modulation Transfer Function

Now let us consider another modification of sum rule:

$$\sum_{(2k-1)|(2s-1)} \mu(2k-1) = \delta_{2s-1,1} = \delta_{s,1} \qquad (1.85)$$

Correspondingly we have

$$G(x) = \sum_{k=1}^{\infty} \frac{F((2k-1)x)}{(2k-1)^n} \Longleftrightarrow$$
$$F(x) = \sum_{k=1}^{\infty} \frac{\mu(2k-1)G((2k-1)x)}{(2k-1)^n} \qquad (1.86)$$

and

$$G(x) = \sum_{k=1}^{\infty} \frac{(-1)^{k-1}F((2k-1)x)}{(2k-1)^n} \Longleftrightarrow$$
$$F(x) = \sum_{k=1}^{\infty} \frac{(-1)^{k-1}\mu(2k-1)G((2k-1)x)}{(2k-1)^n} \qquad (1.87)$$

Fig. 1.7 Pafnuty Chebyshev (1821–1894)

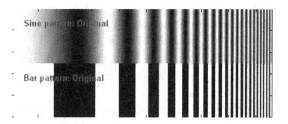

Fig. 1.8 Comparison of sine and bar pattern

The proof is similar as before, with the same parity of $s-1$ and $k+\ell$ when $2s-1 = (2\ell-1)(2k-1)$. Note that the above two Möbius–Chebyshev inverse formulas were proposed by Chebyshev first in 1851[Che1851][8]. Now let us consider the application to the modulation transfer function.

The modulation depth of a grating is defined as the relative variation of brightness

$$M = \frac{I_{\max} - I_{\min}}{I_{\max} + I_{\min}} \qquad (1.88)$$

[8]Chebyshev is considered a founding father of Russian mathematics. Among his well-known students were the prolific mathematicians Dmitry Grave, Aleksandr Korkin, Aleksandr Lyapunov and Andrei Markov. According to the Mathematics Genealogy Project, Chebyshev has about 5,000 mathematical "descendants". A physical handicap (of unknown cause) affected Chebyshev's adolescence and development. From childhood, he limped and walked with a stick and so his parents abandoned the idea of his becoming an officer in the family tradition. His disability prevented his playing many children's games and he devoted himself instead to the passion of his life, working on his math. [From Wikipedia]

and defining the modulation transfer function (MTF) as the ratio of modulation depth in the image to that in the object

$$\mathbf{T}(\nu) = \frac{M_{\text{ima}}(\nu)}{M_{\text{obj}}(\nu)} \tag{1.89}$$

where M_{ima} and M_{obj} are modulation depth of image and object respectively, the modulation function represents the variation of modulation depth, or the aberration of the image, or the decrease of contrast during image formation. According to the standard, the modulation function should set a sine grating to form the image, then to measure the modulation depth before and after the sine grating. As shown in Fig. 1.8, when we plot modulation transfer against spatial frequency, we obtain the MTF, generally a decreasing function of spatial frequency. However, the accuracy of sine graticule is hard to realize. In practice we make a rectangular graticule to measure the maximum and minimum brightness (or the ratio of modulation depth) before and after image formation. This is called 'contrast MTF', denoted as $T_{\text{rec}}(\nu)$[Col54; Mai79], and

$$T_{\text{rec}}(\nu) = \frac{M_{\text{ima}}^{\text{rec}}(\nu)}{M_{\text{obj}}^{\text{rec}}(\nu)} \tag{1.90}$$

Note that the definition of "spatial period" is still certain, but "spatial frequency" has lost certainty in this case. Now one needs to extract the unmeasurable $T(\nu)$ from the measurable $T_{\text{rec}}(\nu)$. Suppose that the brightness distribution of a rectangular grating with period of $(1/\nu)$ is as

$$I(x) = \frac{1}{2} + \frac{2}{\pi}[\sin 2\pi\nu x + \frac{1}{3}\sin 2\pi \cdot 3\nu x + \frac{1}{5}\sin 2\pi \cdot 5\nu x + \cdots] \tag{1.91}$$

and the MTF of the image formation system is $\mathbf{T}(\nu)$, then the brightness distribution $I'(x)$ of the rectangular grating can be evaluated by

$$I'(x) = \frac{1}{2} + \frac{2}{\pi}[\mathbf{T}(\nu)\sin 2\pi\nu x + \frac{1}{3}\mathbf{T}(3\nu)\sin 2\pi \cdot 3\nu x + \\ + \frac{1}{5}\mathbf{T}(5\nu)\sin 2\pi \cdot 5\nu x + \cdots] \tag{1.92}$$

The maximum brightness of image is located at $2\pi\nu x = \pi/2$, correspondingly,

$$I'_{\max} = \frac{1}{2} + \frac{2}{\pi}[\mathbf{T}(\nu) - \frac{1}{3}\mathbf{T}(3\nu) + \frac{1}{5}\mathbf{T}(5\nu) - \cdots]$$

and the minimum at $2\pi\nu x = 3\pi/2$ corresponds to

$$I'_{\min} = \frac{1}{2} - \frac{2}{\pi}[\mathbf{T}(\nu) - \frac{1}{3}\mathbf{T}(3\nu) + \frac{1}{5}\mathbf{T}(5\nu) - \cdots]$$

Thus we have
$$I'_{max} - I'_{min} = \frac{4}{\pi}[\mathbf{T}(\nu) - \frac{1}{3}\mathbf{T}(3\nu) + \frac{1}{5}\mathbf{T}(5\nu) - \cdots]$$
and
$$I'_{max} + I'_{min} = 1$$
Therefore, it is given by $M^{rec}_{obj}(\nu) = 1$ and
$$M^{rec}_{ima}(\nu) = \frac{4}{\pi}[\mathbf{T}(\nu) - \frac{1}{3}\mathbf{T}(3\nu) + \frac{1}{5}\mathbf{T}(5\nu) - \cdots]$$
or
$$T_{rec}(\nu) = \frac{4}{\pi}[\mathbf{T}(\nu) - \frac{1}{3}\mathbf{T}(3\nu) + \frac{1}{5}\mathbf{T}(5\nu) - \cdots] \quad (1.93)$$

This shows how to obtain the contrast MTF from the MTF. Now we are interested in the inverse problem. The conventional way to deduce this formula is quite different[Mai79; Bor2001]. They use the elimination method.

For eliminating the term of $\mathbf{T}(3\nu)$ in the right hand of Eq. (1.93), to replace the variable ν by 3ν, and multiplying both sides by $(1/3)$, it follows that
$$\frac{1}{3}T_{rec}(3\nu) = \frac{4}{3\pi}[\mathbf{T}(3\nu) - \frac{1}{3}\mathbf{T}(9\nu) + \frac{1}{5}\mathbf{T}(15\nu) - \cdots]$$

Similarly, for eliminating the term of $\mathbf{T}(5\nu)$ in the right hand of Eq. (1.93), to replace the variable ν by 5ν, and multiplying both sides by $(1/5)$, then
$$\frac{1}{5}T_{rec}(5\nu) = \frac{4}{5\pi}[\mathbf{T}(5\nu) - \frac{1}{3}\mathbf{T}(15\nu) + \frac{1}{5}\mathbf{T}(25\nu) - \cdots]$$

and so on, then constantly to add and subtract these expressions, we have
$$T_{rec}(\nu) + \frac{1}{3}T_{rec}(3\nu) - \frac{1}{5}T_{rec}(5\nu) + \frac{1}{7}T_{rec}(7\nu) + \cdots = \frac{4}{\pi}\mathbf{T}(\nu)$$
or
$$\mathbf{T}(\nu) = \frac{\pi}{4}[T_{rec}(\nu) + \frac{1}{3}T_{rec}(3\nu) - \frac{1}{5}T_{rec}(5\nu) + \frac{1}{7}T_{rec}(7\nu)$$
$$+ \frac{1}{11}T_{rec}(11\nu) - \frac{1}{13}T_{rec}(13\nu) - \cdots + \frac{B_n}{n}T_{rec}(n\nu) + \cdots] \quad (1.94)$$

where B_n is
$$B_n = \begin{cases} (-1)^m(-1)^{(n-1)/2} & if \quad r = m \\ 0 & if \quad r < m \end{cases} \quad (1.95)$$

in which n is of odd integers, and m the total number of the prime factors of n, r the total number of distinct prime factors of n. $r = m$ indicates that n is a product of r distinct prime; $r < m$ indicates that n includes

Table 1.2 Coefficients of modulation transfer function

k	1	2	3	4	5	6	7	8	9	10
$\mu(2k-1)$	1	-1	-1	-1	0	-1	-1	1	-1	-1
$(-1)^{k+1}\mu(2k-1)$	1	1	-1	1	0	1	-1	-1	-1	1
k	11	12	13	14	15	16	17	18	19	20
$\mu(2k-1)$	-1	0	-1	1	1	0	-1	0	-1	0
$(-1)^{k+1}\mu(2k-1)$	-1	0	-1	-1	1	0	-1	0	-1	0

repeated factors. For example, $n = 15$ corresponds to $r = m = 2$; $r = 12$ corresponds to $m = 3$ and $r = 2$. For the sawtooth wave, the similar result was obtained[Col54]. This deduction appeared about right after the end of World War II, and it is still fashionable now[Bor2001]. But now this method is not recommended since we have the much simpler Möbius inversion technique.

If we use Chebyshev's Möbius inversion theorem Eq. (1.86), it is simply given that

$$\mathbf{T}(\nu) = \frac{\pi}{4} \sum_{k=1}^{\infty} \frac{(-1)^{k+1}\mu(2k-1)}{2k-1} T_{\text{rec}}((2k-1)\nu)$$

$$= \frac{\pi}{4}[T_{\text{rec}}(\nu) + \frac{1}{3}T_{\text{rec}}(3\nu) - \frac{1}{5}T_{\text{rec}}(5\nu)$$

$$+ \frac{1}{7}T_{\text{rec}}(7\nu) + \frac{1}{11}T_{\text{rec}}(11\nu) - \cdots] \quad (1.96)$$

Comparing to Eq. (1.96), there is a explicit relation between B_{2k-1} and $\mu(2k-1)$

$$B_{2k-1} = \mu(2k-1)(-1)^{k+1}$$
$$= \begin{cases} (-1)^s(-1)^{k+1} & \text{if } 2k-1 = \prod p_i, \ p_i \neq p_j \text{ for } i \neq j \\ 0 & \text{if } p^2|2k-1 \end{cases} \quad (1.97)$$

shown in Table 1.2. The Möbius technique can be easily used on other gratings.

1.6 Witten Index and Möbius Function

D. Spector first pointed out the connection between Möbius function $\mu(n)$ and Witten operator $(-1)^F$ in the supersymmetric quantum field theory

(QFT)[Spe90], thus quite a few Möbius formulas have new interpretations in the framework of the supersymmetric QFT. Conventionally, in a certain supersymmetric system there is a countable infinity of bosonic creation operator $b_1^\dagger, b_2^\dagger, b_3^\dagger, \ldots$ In principle, the state of a single particle can be expressed as

$$b_j^\dagger |vacuum\rangle \equiv b_j^\dagger |0\rangle \tag{1.98}$$

or

$$|1\rangle,\ |2\rangle,\ |3\rangle,\ |4\rangle,\ldots \tag{1.99}$$

and a state of a many-particle system should be expressed as

$$\prod_{j=1}^{k}(b_j^\dagger)^{\alpha_j}|0\rangle \tag{1.100}$$

where both j and α_j are positive integers. There are infinite numbers of creation operators and the corresponding states, thus the expression of a many-particle state is quite complicated. For example,

$$(b_3^\dagger)(b_2^\dagger)^2(b_1^\dagger)|0\rangle \tag{1.101}$$

and

$$(b_3^\dagger)(b_2^\dagger)(b_1^\dagger)^2|0\rangle \tag{1.102}$$

both are 4-particle states.

Now let us designate these states by primes; this is just the Gödel numbering in mathematics[9]. According to Gödel, the single particle states in Eq. (1.99) can be expressed primes as

$$|p_1\rangle = |2\rangle,\ |p_2\rangle = |3\rangle,\ |p_3\rangle = |5\rangle,\ |p_4\rangle = |7\rangle,\ |p_5\rangle = |11\rangle,\ldots \tag{1.103}$$

and the vacuum state can be expressed as

$$|p_0\rangle = |1\rangle = |vacuum\rangle \tag{1.104}$$

Obviously, there are many integers besides the primes. What are their usages? In fact, according to the theorem on unique factorization of a natural number, we have

$$N \equiv (p_{i_1})^{\alpha_1}(p_{i_2})^{\alpha_2}\ldots(p_{i_s})^{\alpha_s} \tag{1.105}$$

---

[9]Kurt Gödel was an Austrian-American logician, mathematician and philosopher. One of the most significant logicians of all time, Gödel made an immense impact upon scientific and philosophical thinking in the 20th century, a time when many, such as Bertrand Russell, A.N. Whitehead and David Hilbert, were pioneering the use of logic and set theory to understand the foundations of mathematics. In 1933, Gödel first travelled to the U.S., where he met Albert Einstein in Princeton, who became a lifelong friend. From Wikipedia.

Fig. 1.9 Albert Einstein (1879–1955) and Kurt Gödel (1906–1978)

Similarly,

$$|N\rangle \equiv \prod_{j=1}^{s} [b_{p_j}^{\dagger}]^{\alpha_j} |vacuum\rangle \qquad (1.106)$$

In this new representation, a prime number can represent a single-particle state as in Eq. (1.106), and a composite number can represent a many-particle state, for example, $|90\rangle$ and $|60\rangle$ can represent the many-particle states of Eqs. (1.101) and (1.102). In other words, any positive integer N represents a state of particle

$$\begin{cases} single-particle\ state & when\ N\ is\ a\ prime\ number, \\ many-particle\ state & when\ N\ is\ a\ composite\ number. \end{cases}$$

For convenience, in the case of use of Gödel numbering, we denote

$$|p_1\rangle = b_1^{\dagger}|1\rangle,\ |p_2\rangle = b_2^{\dagger}|1\rangle,\ |p_3\rangle = b_3^{\dagger}|1\rangle,\ |p_4\rangle = b_4^{\dagger}|1\rangle,\ ... \qquad (1.107)$$

Now let consider a system including both bosons and fermions with Gödel numbering. Suppose that in a certain supersymmetric system there is a countable infinity of boson creation operators $b_1^{\dagger}, b_2^{\dagger}, b_3^{\dagger}, ...$ and ordered fermion creation operators $f_1^{\dagger}, f_2^{\dagger}, f_3^{\dagger}, ...$ By Gödel numbering, a one-to-one correspondence between these operators and prime numbers is established such that both $b_i^{\dagger}$ and $f_i^{\dagger}$ correspond to the i-th prime p_i, $i = 1, 2, 3, ...$ Thus, the boson operator $\prod_{j=1}^{r}\{b_j^{\dagger}\}^{\alpha_j}$ naturally corresponds to the number $\prod_{j=1}^{r} p_j^{\alpha_j}$, where j and α_j are integers, and the fermion operator $\prod_{j=1}^{r}(f_j^{\dagger})^{\beta_j}$

to the number $\prod_{j=1}^{r} p_j^{\beta_j}$, where β_j can only take values of -1 and 1, due to their anti-commutativity or Pauli's exclusion principle. In other words, the former associates with all positive integers and the latter with all square free integers. As operators $b_i^\dagger$, $f_i^\dagger$ act on the vacuum state $|1>$ of the theory in all possible ways, all eigenstate of the Hamiltonian H of this supersymmetric QFT are created in a Fock space. We denote the normalized states as follows:

$$\prod_{j=1}^{r}(b_j^\dagger)^{\alpha_j}(f_j^\dagger)^{\beta_j}|0> = |N,d\rangle \qquad (1.108)$$

where

$$N = \prod_{j=1}^{r}(p_j^\dagger)^{\alpha_j+\beta_j} \qquad (1.109)$$

and

$$d = \prod_{j=1}^{r}(p_{i_j}^\dagger)^{\beta_j} \qquad (1.110)$$

The number N indicates the total excitation of the state $|N,d\rangle$ above the vacuum, the d the fermion part, and the complete $|N,d\rangle$ together become a complete set of states of the theory. By using these notations, d is evidently a factor of N, and we can calculate the Witten index Δ of this system, which is defined as

$$\Delta = Tr(-1)^F \qquad (1.111)$$

or

$$\Delta = Tr[(-1)^F e^{-\beta H}], \ 0 < \beta < 1 \qquad (1.112)$$

by regularization, it has a distinct meaning in the supersymmetric QFT since we have

$$\Delta = Tr[(-1)^F e^{-\beta H}] = \sum_{N=1}^{\infty}\sideset{}{'}\sum_{d|N} \langle N,d|(-1)^F e^{-\beta H}|N,d\rangle$$

$$= \sum_{N=1}^{\infty}\sideset{}{'}\sum_{d|N} \langle N,d|e^{-\beta H}|N,d\rangle\langle N,d|(-1)^F|N,d\rangle \qquad (1.113)$$

where $\sum'$ represents the summation that only runs over the square-free numbers, and $d|N$ means that d runs over all the factors of N including 1 and N. It should be mentioned here that there are two important properties of the states $|N,d\rangle$. First, the energy of $|N,d\rangle$ is a function of N and independent of d; the expected value of the operator $(-1)^F$ above $|N,d\rangle$ depends only on d, and is independent of N. Furthermore, the operator $(-1)^F$, as acted on the state $|N,d\rangle$, is equal to $+1$ or -1 depending on even or odd number of fermion excitation in $|N,d\rangle$. For a system with unbroken supersymmetry ($\Delta = 1$) and a mass gap, we define

$$\mu(d) = \left\{ \begin{array}{ll} \langle N,d|(-1)^F|N,d\rangle, & \text{if } p^2 \nmid d \\ 0, & \text{otherwise.} \end{array} \right\} \quad (1.114)$$

then $\mu(d)$ is just the Möbius function defined as Eq. (1.19), $p^2 \nmid d$ means d is square free. Thus, the sum rule of $\mu(d)$ can be understood in the supersymmetric language: the set of states with a given value of $N > 1$ gives a vanished contribution to the Witten index. For details readers may refer to Spector[Spe90].

1.7 Cesàro–Möbius Inversion Formula

Cesàro proposed a similar multiplicative Möbius inverse formula in 1885. Omitting the issues of convergence, the Cesàro-Möbius inversion formula states as follows.

Theorem 1.16. If for any integers m and n, there exists

$$g_m(g_n(x)) = g_{mn}(x) \quad (1.115)$$

and

$$F(x) = \sum_{n=1}^{\infty} a(n) f(g_n(x)) \iff f(x) = \sum_{n=1}^{\infty} a^{-1}(n) F(g_n(x)) \quad (1.116)$$

with

$$\sum_{n|k} a^{-1}(n) a(\frac{k}{n}) = \delta_{k,1} \quad (1.117)$$

Proof. In fact,

$$\sum_{n=1}^{\infty} a^{-1}(n) F(g_n(x)) = \sum_{n=1}^{\infty} a^{-1}(n) \sum_{m=1}^{\infty} a(m) f(g_m[g_n(x)])$$

$$= \sum_{n=1}^{\infty} [\sum_{n|k} a^{-1}(n) a(\frac{k}{n})] f(g_k(x)) = \sum_{n=1}^{\infty} \delta_{k,1} f(g_k(x))$$

$$= f(g_1(x)) = f(x)$$

□

Fig. 1.10   Ernesto Cesàro (1859–1906)

From the first double sum to the second one, it is assumed that

$$\sum_{n=1}^{\infty}\sum_{m=1}^{\infty} |a^{-1}(n)a(m)f(g_m[g_n(x)])| < \infty$$

Equation (1.116) is also called Möbius-Cesàro inverse formula[10]. Several interesting examples are listed as follows.

$$g_n(x) = \frac{x}{1 - x \ln n} \qquad (1.118)$$

Noted that

$$g_1(x) = \frac{x}{1 - x \ln 1} = \frac{x}{1 - x \cdot 0} = x$$

and

$$g_m(g_n(x)) = g_m\left(\frac{x}{1 - x \ln n}\right)$$
$$= \frac{\frac{x}{1 - x \ln n}}{1 - \frac{x}{1 - x \ln n} \ln m}$$
$$= \frac{x}{1 - x \ln n - x \ln m}$$
$$= \frac{x}{1 - x \ln mn} = g_{mn}(x)$$

Another example is

$$g_n(x) = \frac{xn}{1 - kx[n-1]} \qquad (1.119)$$

---

[10]Cesàro was an Italian mathematician. In his work on mathematical physics he was a staunch follower of Maxwell. This helped to spread Maxwell's ideas to the Continent which was important since, although it is hard to believe this now, it took a long time for scientists to realize the importance of his theories. [From MacTutor biography]

where k is a constant. And
$$g_1(x) = \frac{x}{1 - kx[1-1]} = \frac{x}{1 - kx \cdot 0} = x$$
and
$$g_m(g_n(x)) = g_m(\frac{xn}{1 - kx[n-1]})$$
$$= \frac{\frac{xn}{1-kx[n-1]}m}{1 - k\frac{xn}{1-kx[n-1]}[m-1]}$$
$$= \frac{xnm}{1 - kx[n-1] - kxn[m-1]}$$
$$= \frac{xnm}{1 - kx[mn - 1]} = g_{mn}(x)$$
In fact, the above two examples belong to a more general situation as
$$g_n(x) = F^{-1}(\ln n + F(x)) \qquad (1.120)$$
Noted that F(x) is an arbitrary function, and $F^{-1}F(x)=x$, we have
$$g_1(x) = F^{-1}(\ln n + F(x))|_{n=1} = F^{-1}(F(x)) = x$$
and
$$g_m(g_n(x)) = g_m(F^{-1}[\ln n + F(x)])$$
$$= F^{-1}(\ln m + F\{F^{-1}(\ln n + F(x))\})$$
$$= F^{-1}(\ln m + [\ln n + F(x)])$$
$$= F^{-1}[\ln mn + F(x)] = g_{mn}(x)$$
Here are some examples:
$$g_n(x) = \sqrt{\ln n + x^2} \qquad (1.121a)$$
$$g_n(x) = (\ln n + \sqrt{x})^2 \qquad (1.121b)$$
$$g_n(x) = \ln(\ln n + e^x) \qquad (1.121c)$$
$$g_n(x) = \exp(\ln n + \ln x) \qquad (1.121d)$$
$$g_n(x) = \sin(\ln n + \sin^{-1} x) \qquad (1.121e)$$
$$g_n(x) = \sin^{-1}(\ln n + \sin x) \qquad (1.121f)$$
Obviously, we might get more generalized examples, if $\ln n$ in Eq. (1.136) is replaced by $\ln b(n)$, where $b(m)b(n) = b(mn)$.

Let us denote $g_n(x)$ as $g(x, n)$, now we are going to prove a theorem to explain Eq. (1.120).

Theorem 1.17.
$$g(g(x,n), m) = g(x, mn) \Rightarrow g_n(x) = F^{-1}(\ln n + F(x)) \qquad (1.122)$$

The proof is given as follows. Introducing a function h

$$h(x,t) = g(x, e^t) = g(x, n) \qquad (1.123)$$

and assume that the function $g(x, n)$ has derivative with respect to the "virtual continuous variable" n. From $g(g(x, n), m) = g(x, mn)$, we have

$$h(h(x,t), \Delta t) = h(x, t+ \Delta t) \qquad (1.124)$$

with

$$\frac{\partial}{\partial t} h(x,t) = f(h(x,t)) \qquad (1.125)$$

Note that $g(x, 1) = x$ and $n = 1$ corresponds to $t = 0$, thus

$$h(x, 0) = x \qquad (1.126)$$

From Eq. (1.125), it follows that

$$\left\{ \frac{\partial}{\partial t} h(x,t) \right\} |_{t=0} = f(h(x,0)) = f(x) \qquad (1.127)$$

and

$$\frac{dx}{dt} = f(x) \qquad (1.128)$$

or

$$\frac{dx}{f(x)} = dt \qquad (1.129)$$

This expression is not only for Eq. (1.126) but also for Eq. (1.125). In other words, x can represent $h(x, t)$ in Eq. (1.129) no matter what value t is. Therefore, we can define a new function $F(x)$ such that

$$F(x) \equiv \int \frac{dx}{f(x)} = t + const \qquad (1.130)$$

and

$$h(x,t) = F^{-1}\{F(h(x,t))\} = F^{-1}(t + const) \qquad (1.131)$$

In order to determine the constant, we consider the case of $t = 0$ as

$$x = h(x, 0) = F^{-1}(t + const) \qquad (1.132)$$

That is,

$$const = F(x) \qquad (1.133)$$

since this "constant" is only time-invariant. Combining Eqs. (1.131) and (1.133), we have

$$h(x,t) = F^{-1}\{F(h(x,t))\} = F^{-1}(t + F(x)) \qquad (1.134)$$

This is equivalent to

$$g(x,n) \equiv g_n(x) \equiv h(x,t)|_{t=\ln n} = F^{-1}(\ln n + const) \qquad (1.135)$$

or

$$g_n(x) = F^{-1}(\ln n + F(x)) \qquad (1.136)$$

This formula will be used for the design of some new modulation systems, and one of its modifications will be applied to interface study.

There is a modified Cesàro formula, which can be expressed as

$$H(g_n(x)) = \sum_{d|n} r(d) h(g_{\frac{n}{d}}(x)) \Leftrightarrow h(g_n(x)) = \sum_{d|n} r^{-1}(d) H(g_{\frac{n}{d}}(x)) \qquad (1.137)$$

Proof.

$$\sum_{d|n} r^{-1}(d) H(g_{\frac{n}{d}}(x))$$
$$= \sum_{d|n} r^{-1}(d) \sum_{d'|d} r(d') H(g_{d'}[g_{\frac{n}{d}}(x)])$$
$$= \sum_{d|n} \sum_{d'|d} r^{-1}(d) r(d') H([g_{\frac{n/d}{d'}}(x)])$$
$$= \sum_{k|n} \sum_{d|k} r^{-1}(\frac{k}{d}) r(d) H([g_{\frac{n}{k}}(x)])$$
$$= \sum_{k|n} \delta_{k,1} H([g_{\frac{n}{k}}(x)]) = H([g_n(x)]$$

$\square$

1.8 Unification of Eqs. (1.20) and (1.47)

From a historical point of view, the Möbius series inversion formula Eq. (1.20) is even more original and classical, and the classical Möbius inverse formula Eq. (1.47) in number theory is the result of purification (or polishing) of the series inversion formula in order to avoid the divergence.

This section introduces a theorem which formally unifies Eqs. (1.20) and (1.47) as follows.

Theorem 1.18.

$$g_n(x) = \sum_{d|n} f_{\frac{n}{d}}(dx) \Leftrightarrow f_n(x) = \sum_{d|n} \mu(d) g_{\frac{n}{d}}(dx) \qquad (1.138)$$

Proof.

$$\sum_{d|n} \mu(d) g_{\frac{n}{d}}(dx) = \sum_{d|n} \mu(d) \sum_{d'|\frac{n}{d}} f_{\frac{n}{d'd}}(d'dx)$$

$$\overset{k=dd'}{=} \sum_{k|n} \{\sum_{d|k} \mu(d)\} f_{\frac{n}{k}}(kx)$$

$$= \sum_{k|n} \delta_{k,1} f_{\frac{n}{k}}(kx) = f_n(x)$$

□

(1) Case $n = 0$. Note that $d|0$ for an arbitrary integer d, it implies that

$$\sum_{d|0} \equiv \sum_{d=1}^{\infty}$$

so that

$$g_0(x) = \sum_{d=1}^{\infty} f_0(dx) \Leftrightarrow f_0(x) = \sum_{d=1}^{\infty} \mu(d) g_0(dx)$$

Omitting the constant parameter 0, this relation can be rewritten as

$$G(x) = \sum_{d=1}^{\infty} F(dx) \Leftrightarrow F(x) = \sum_{d=1}^{\infty} \mu(d) G(dx)$$

This is right the Möbius series inversion theorem (1.20).

(2) Case $x = 0$. Now we have

$$g_n(0) = \sum_{d|n} f_{\frac{n}{d}}(0) \Leftrightarrow f_n(0) = \sum_{d|n} \mu(d) g_{\frac{n}{d}}(0)$$

To omit the definitive parameter $x = 0$, and consider the subscript as variable, this is just the same as the classical Möbius inverse formula Eq. (1.20).

It is concluded that the new theorem unifies the Eqs. (1.20) and (1.47).

Similarly, a generalized Cesàro-Möbius inversion formula can be expressed as

$$H(n,x) = H(n, g_1(x)) = \sum_{d|n} r(d) h(n/d, g_d(x)) \Leftrightarrow$$

$$h(n,x) = \sum_{d|n} r^{-1}(d) H(n/d, g_d(x)) \qquad (1.139)$$

the proof is omitted. Obviously, the case of $n = 0$ corresponds to Eq. (1.116), the case of $x = 0$ corresponds to Eq. (1.63).

1.9 Summary

Chapter 1 has presented the basic knowledge for Möbius inversion formulas with some simple applications in physics. This chapter addressed several specific topics.

(1) For convenience, a formal unification of the Möbius inverse formula and the Möbius series inversion formula was presented.

(2) The Chebyshev–Möbius series inversion formulas and applications thereof.

(3) A detailed demonstration of the Möbius–Cesàro inversion formula.

(4) The importance of the varying sum roles.

(5) A historic note on Möbius's contribution in 1832.

This chapter also addressed some fragmentary applications of the Möbius serious inversion formula and its various modifications. Beginning with the next chapter, we will see some systematic applications of the Möbius inversion method. Note that the relevant convergence problem is very interesting but omitted quite often in this book since the world of infinite series is just too big and complex[Kno28; Sza47]. In physics, the situation is mostly twofold. One aim is to evaluate the convergence or divergence properties of some existing functional series, and the other is to design some functional forms for unknown physical properties under some restriction and controlling. Ironically, the discovery of the Möbius inversion was accidental, caused by an incomplete mathematical procedure. Finally, let us quote a sentence from Poincaré, " The pure mathematician who should forget the existence of the exterior would be like a painter who knows how to harmoniously combine colors and forms, but who lacked models. His creative power would soon be exhausted."

1.10 Supplement – the Seminal Paper of Möbius

The original paper on Möbius inverse formula in 1832 can be found in the internet[11].

---

[11] University of Michigan Historical Math Collection online service.

On One Special Type of Inversion of Series, by A.F. Möbius, Professor in Leipzig, 1832

The famous problem of inversion of a series is known to consist in that, when a function of a variable is defined by a series of increasing powers, one requires the inversion of that variable itself, or any other function of it, to be expressed through a series of progressing powers of that function. As is well known, no small analytical skills are needed to find the law that relates the coefficients of the second series to those of the first one. Incomparably simpler is the following problem.

Let $f(x)$ be a function of a variable x given as an ordered series of powers of x:

1. $$f(x) = a_1 x + a_2 x^2 + a_3 x^3 + a_4 x^4 + \cdots \qquad (1.140)$$

The variable x is to be expressed not by means of powers of f(x) but by means of functions f of the powers of x, as represented by the following series:

2. $$x = b_1 f(x) + b_2 f(x^2) + b_3 f(x^3) + b_4 f(x^4) + \cdots \qquad (1.141)$$

Whereas this problem can be posed either in the context of the difficulties of its solution or in the context of its uses in the aforementioned problem of series inversion, in which neither the values of $f(x), f(x^2), f(x^3), \ldots$ nor the value of x can be calculated through formula (2.), thus leaving real purpose of the inversion problem unattained, the solution of this problem will nevertheless lead to many results not unimportant for the theory of series as well as the theory of combinatorics.

The essence of our problem is this: to express the coefficients $b_1, b_2, b_3, \ldots$ of series (2.) as functions of the coefficients $a_1, a_2, a_3, \ldots$; and this can be done through the following very easy calculations. From (1.) it follows that

$$f(x^2) = a_1 x^2 + a_2 x^4 + a_3 x^6 + a_4 x^8 + \cdots$$
$$f(x^3) = a_1 x^3 + a_2 x^6 + a_3 x^9 + \cdots$$
$$f(x^4) = a_1 x^4 + a_2 x^8 + a_3 x^{12} + \cdots$$
$$f(x^5) = a_1 x^5 + a_2 x^{10} + \cdots$$
$$f(x^6) = a_1 x^6 + a_2 x^{12} + \cdots$$
$$\cdots\cdots$$

Substituting these values of $f(x), f(x^2), f(x^3), \ldots$ and of $f(x)$ directly into (2.), one obtains:

$$x = a_1 b_1 x + \begin{vmatrix} +a_2 b_1 \\ +a_1 b_2 \end{vmatrix} x^2 + \begin{vmatrix} +a_3 b_1 \\ +a_1 b_3 \end{vmatrix} x^3 + \begin{vmatrix} +a_4 b_1 \\ +a_2 b_2 \\ +a_1 b_4 \end{vmatrix} x^4 + \begin{vmatrix} +a_5 b_1 \\ +a_1 b_5 \end{vmatrix} x^5 + \begin{vmatrix} +a_6 b_1 \\ +a_3 b_2 \\ +a_2 b_3 \\ +a_1 b_6 \end{vmatrix} x^6 + \cdots$$
(1.142)

The rule of progression of the coefficients of this series is obvious. If just the coefficients of x^m is to be determined, one needs to decompose the number m in all possible ways into two positive factors. Each of these products then gives one term of the coefficient sought, as a result of which one takes two factors of the products as the indices of the mutually multiplied a and b.

Since the equation just obtained must apply for each value of x, we have

3. $\begin{cases} a_1 b_1 = 1 \\ a_2 b_1 + a_1 b_2 = 0 \\ a_3 b_1 + a_1 b_3 = 0 \\ a_4 b_1 + a_2 b_2 + a_1 b_4 = 0 \\ a_5 b_1 + a_1 b_5 = 0 \\ a_6 b_1 + a_3 b_2 + a_2 b_3 + a_1 b_6 = 0 \\ \ldots \end{cases}$ (1.143)

through which each b can be calculated with the help of the preceding b.

To find, in this manner, the individual b independently of each other,

Fig. 1.11 A stamp with a Möbius ring

one may set for the sake of simplicity $a_1 = 1$, and it ensues

$$
4. \begin{cases} b_1 = 1 \\ b_2 = -a_2 \\ b_3 = -a_3 \\ b_4 = -a_4 + a_2 a_2 \\ b_5 = -a_5 \\ b_6 = -a_6 + a_3 a_2 + a_2 a_3 \\ b_7 = -a_7 \\ b_8 = -a_8 + a_4 a_2 + a_2 a_4 - a_2 a_2 a_2 \\ \cdots \end{cases} \qquad (1.144)
$$

Already these limited developments suffice for inferring how also the values of the successive b must be assembled from $a_1, a_2, a_3, \ldots$ Indeed, one decomposes index m in all possible ways into factors, taking m itself as the highest factor but omitting the unity, and regarding each two decompositions that differ only by the order of their factors as different. Or, as can be concisely stated in the language of combinatorics: One constructs all the variations with repetitions for the product m. Each of these variations thus yields one term in the value of b_m, which takes the index m in the elements of variation, and this term receives either a positive or negative sign, depending on whether the number of its elements is even or odd.

So, for instance, the variations of product 12 are:

$$12, \; 2 \times 6, \; 3 \times 4, \; 4 \times 3, \; 6 \times 2, \; 2 \times 2 \times 3, \; 2 \times 3 \times 2, \; 3 \times 2 \times 2$$

and from this

$$b_{12} = -a_{12} + 2 a_2 a_6 + 2 a_3 a_4 - 3 a_2 a_2 a_2$$

To give a very simple example of this new kind of series inversion, we may set

$$a_1 = a_2 = a_3 = \cdots = 1$$

also from (1.)

$$f(x) = x + x^2 + x^3 + \cdots \text{ and from that} f(x) = \frac{x}{1-x}$$

With these values of a, however, according to (4.): $b_1 = 1, b_2 = -1, b_3 = -1, b_4 = 0, b_5 = -1, b_6 = 1, b_7 = -1, b_8 = 0, \cdots$ etc.

One would surely find it very pleasing that the coefficients of this series, when it is continued ever farther, are none of other then $1, 0$, and -1. The basis of this remarkable result, and the law according to which the coefficients $1, 0$ and -1 interchange with each other, can be most easily uncovered with the help of recursion formula (3.)

From the above seminal paper of Möbius, it can be concluded that Möbius did not actually derive the Möbius inverse formula, although he did discover all the necessary ingredients, such as the Möbius functions and the sum rule of Möbius functions, to do so. What Möbius actually studied was the inversion of infinite series, that is Möbius series inversion formula Eq. (1.20), but (1.63). As a student of Gauss, it is not surprising that Möbius solved the series inversion by using Gaussian elimination, expressing the solution in such a sophisticated way. In other words, the rigorousness of this original paper was not perfect, but the creativity is extremely exciting. Even though the formula (1.63) is perfect with various applications in mathematics, the nearly forgotten formula (1.20) has demonstrated its application to multiple discipline in physics.

Chapter 2

Inverse Problems in Boson Systems

2.1 What is an Inverse Problem?

Definition 2.1. Two problems are inverse to each other if the formulation of each of them requires full or partial knowledge of the other.

By this definition[Kir96; Gro99], it is obviously arbitrary which of the two problems we call the direct problem and which the inverse problem. Usually, the problem which has been studied earlier and in more detail is called the direct problem, and the other the inverse problem[Kir96]. Most exercise problems in textbooks are direct problems, and most problems we encounter in research are inverse problems. With the arrival of the Information Age, a variety of inverse problems have been proposed and solved, and many relevant conferences and journals have emerged in recent years.

Now let us consider a direct problem in elementary arithmetic operation: what is 2 to the 5th power? The answer is 32, as

$$2^5 = 32$$

One inverse problem to the above problem is: what is the 5th root of 32? There are five solutions as:

$$32^{1/5} = 2, \text{ or } 2e^{\pm 2\pi i/5}, \text{ or } 2e^{\pm 4\pi i/5}$$

Another inverse problem to the above direct problem is: what is the logarithm of 32 with base 2? This time there are an infinite number of solutions as

$$\log_2 32 = 5 + \frac{2\pi n i}{\ln 2}, \quad n \text{ is an integer}$$

The example above illustrates that one problem may correspond to several inverse problems. In this example, the answer to the direct problem is a

Fig. 2.1 Jacques Hadamard (1865–1963)

real number, and the answers to the inverse problems are extended to complex numbers. Also, the answer is unique for the direct problem, and not unique for either of the inverse problems. In reality, the inverse problem may easily become more complex. For example, one may propose a problem such as: what is the solution of $\sqrt[5.01]{32}$? or how many solutions are there for $\sqrt[5.01]{32}$? Obviously, most inverse problems are ill-posed, which means that their solutions are not unique, or not stable, and sometimes there is no solution at all, so we have to change the statement of the problem or the definition of the solution[1].

Now let us consider the situation in physics. In general, the formal logic in a physics textbook is very important; the order for discussing a problem is very often to obtain effect from cause(hypothesis), for instance, to derive the motion of a planet from the gravitation between the sun and the planet in Newtonian mechanics. However, research work in physics is mostly empirical without a fixed regulation: sometimes to obtain the effect(experiment) from the cause, sometimes to speculate or to guess the cause(premise) from the effect(experimental fact). In those days when people were discussing the problem of how to determine the forces between the sun and planets, Newton speculated the gravitational law from Kepler's three laws. In physics, a problem to obtain effect from cause is called a

---

[1] Hadamard claims that a mathematical model for a physical problem has to be well-posed in the sense that it has the following three properties:
⋄ There exists a solution to the problem (existence).
⋄ There is at most one solution to the problem (uniqueness).
⋄ The solution depends continuously on the data (stability).

Fig. 2.2 (a)John William Strutt (Lord Rayleigh) (1842–1919), (b) Sir James Hopwood Jeans (1877–1946), (c)Wilhelm Wien (1864–1928)

direct problem, and a problem to get cause from effect is called an inverse problem. Obviously, there is no causality in solving inverse problems so there is no rigorously logic and systematic method to solve an inverse problem. In fact, that which is known and which is unknown are changing all the time. In other words, once an inverse problem is solved, it becomes a direct problem and the basis for some new inverse problems.

2.2 Inverse Blackbody Radiation Problem

The direct blackbody radiation problem is to give the radiated electromagnetic power spectrum based on the area-temperature distribution on the surface of a blackbody. The solution to this problem can be expressed by

Planck's law of 1900 as follows[2]

$$W(\nu) = \frac{2h\nu^3}{c^2} \int_0^\infty \frac{a(T)dT}{e^{h\nu/kT} - 1}, \qquad (2.1)$$

where $W(\nu)$ is the total radiation power spectrum of a blackbody with area temperature distribution $a(T)$, ν the frequency, T the temperature. The inverse blackbody radiation problem is to determine the area temperature distribution $a(T)$ from the measured total radiation power spectrum $W(\nu)$ of the surface of a blackbody, such as nuclear reactors, integrated circuits, and superior planets. This is very important in remote-sensing and astrophysics; Bojarski proposed the first formulation for the problem in 1982[Boj82]. He presented a numerical solution by using the Laplace transform with iterative process. Subsequently various authors provided different improvements based on his solution[Boj84; Ham83; Lak84; Kim85; Hun86; Rag87; Sun87; Bev89] [Che90a; Dou92; Dou92a; Li2005].

2.2.1 Bojarski iteration

Introducing two variables: coldness or reciprocal temperature u, and area-coldness function $A(u)$ with the definition as

$$u = \frac{h}{k_B T}, \text{ and } A(u) = \frac{a(h/k_B u)}{u^2} \qquad (2.2)$$

then

$$dT = \frac{-h}{k_B} \frac{du}{u^2} \qquad (2.3)$$

---

[2]Wilhelm Wien was a German physicist who, in 1893, used theories about heat and electromagnetism to deduce Wien's displacement law, which calculates the emission of a blackbody at any temperature from the emission at any one reference temperature. He also formulated an expression for the blackbody radiation which is correct in the photon-gas limit. His arguments were based on the notion of adiabatic invariance, and were instrumental for the formulation of quantum mechanics. [From Wikipedia]

Lord Rayleigh was an experimental physicist. His first researches were mainly mathematical, concerning optics and vibrating systems, but his later work ranged over almost the whole field of physics, covering sound, wave theory, color vision, electrodynamics, electromagnetism, light scattering, flow of liquids, hydrodynamics, density of gases, viscosity, capillarity, elasticity, and photography. He had a fine sense of literary style; every paper he wrote, even on the most abstruse subject, is a model of clearness and simplicity of diction. The 446 papers reprinted in his collected works clearly show his capacity for understanding everything just a little more deeply than anyone else. [From Nobelprize.org].

Jeans was an English mathematician, astronomer, and physicist. Jeans also helped to discover the Rayleigh–Jeans law, which relates the energy density of blackbody radiation to the temperature of the emission source. $f(\lambda) = 8\pi c k \frac{T}{\lambda^4}$. [From Wikipedia]

(a) (b)

Fig. 2.3 Image of Max Planck (1858–1947): (a) on a stamp, and (b) in 1901. He said that this radiation law, whenever it is found, will be independent of special bodies and substances and will retain its importance for all times and cultures, even for non-terrestrial and non-human ones. After the great discovery, Planck appeared drained of vitality, exhausted after years of trying to refute his own revolutionary ideas about matter and radiation.

From this, *Bojarski* obtained

$$W(\nu) = \frac{2h^2\nu^3}{k_B c^2} \int_0^\infty \frac{a(h/k_B u)du}{u^2(e^{u\nu} - 1)}. \tag{2.4}$$

By using Taylor expansion, one obtains

$$W(\nu) = \frac{2h^2\nu^3}{k_B c^2} \int_0^\infty \sum_{n=1}^\infty e^{-nu\nu} A(u) du$$

$$= \frac{2h^2\nu^3}{k_B c^2} \int_0^\infty e^{-u\nu} \sum_{n=1}^\infty [\frac{A(u/n)}{n}] du \tag{2.5}$$

Define a function as

$$f(u) = \sum_{n=1}^\infty \frac{A(u/n)}{n} \tag{2.6}$$

then we have a relation with Laplace transform

$$\frac{k_B c^2}{2h^2} \frac{W(\nu)}{\nu^3} = \int_0^\infty f(u) e^{-u\nu} du = \mathcal{L}[f(u); u \to \nu]. \tag{2.7}$$

From this,

$$f(u) = \mathcal{L}^{-1}[\frac{k_B c^2}{2h^2} \frac{W(\nu)}{\nu^3}; \nu \to u] \tag{2.8}$$

where $\mathcal{L}$ and $\mathcal{L}^{-1}$ represent Laplace operator and corresponding inverse operator respectively. The remaining problem is to extract the $A(u)$ through $f(u)$. For designing the iteration process, let

$$A_1(u) = f(u) \text{ and } A_{m+1}(u) = f(u) - \sum_{n=2}^{\infty} \frac{A_m(\frac{u}{n})}{n}. \tag{2.9}$$

Therefore,

$$\lim_{m \to \infty} A_m(u) = A(u). \tag{2.10}$$

That is the Bojarski iteration.

2.2.2 The Möbius inversion for the inverse blackbody radiation

During the nine years after Bojarski proposed the inverse blackbody radiation problem, there were many suggestions on this subject in IEEE[Lak84; Kim85; Hun86]. Here we introduce a very simple and rigorous method which almost terminated this discussion. In fact, Eq. (2.4) can be expressed simply as

$$W(\nu) = \frac{2h^2\nu^3}{k_B c^2} \int_0^{\infty} \frac{a(h/k_B u)}{u^2} \sum_{n=1}^{\infty} e^{-n u \nu} du$$

$$= \frac{2h^2\nu^3}{k_B c^2} \sum_{n=1}^{\infty} \mathcal{L}[u^{-2} a(h/k_B u); u \to n\nu]$$

Based on Möbius series inversion formula, it is given by

$$\mathcal{L}[u^{-2} a(h/k_B u); u \to \nu] = \frac{k_B c^2}{2h^2 \nu^3} \sum_{n=1}^{\infty} \mu(n) \frac{W(n\nu)}{(n\nu)^3} \tag{2.11}$$

the solution thus can be written as

$$u^{-2} a(h/k_B u) = \frac{k_B c^2}{2h^2} \sum_{n=1}^{\infty} \frac{\mu(n)}{n^3} \mathcal{L}^{-1}[\frac{W(n\nu)}{\nu^3}; \nu \to u] \tag{2.12}$$

or

Theorem 2.1.

$$a(T) = \frac{c^2}{2k_B T^2} \sum_{n=1}^{\infty} \frac{\mu(n)}{n^3} \mathcal{L}^{-1}[\frac{W(n\nu)}{\nu^3}; \nu \to u]|_{u=h/k_B T}. \tag{2.13}$$

Fig. 2.4 Ludwig Boltzmann (1844–1906) and Satyendra Nath Bose (1894–1974)

This solution for the inverse blackbody radiation is concise and rigorous, and has been called Chen's inversion formula [Hug90; Nin92; Xie91] for temperature distribution of distorted black holes[Ros93; Ros93a], IRAS data analysis for interstellar dust temperature distribution[Xie91; Xie93], temperature distribution of accretion disks in active galactic nuclei[Wan96] and dust emission[Bet2007]. The details of these applications can be found in Li and Goldsmith[Li99][3].

2.3 Inverse Heat Capacity Problem

2.3.1 *Historical background*

When establishing the thermodynamics for a system, the key point is to determine the energy spectrum of the quasi-particle based on the Hamiltonian operator. Once the energy spectrum is known, all the thermodynamic quantities can be given by integration. The corresponding inverse problem

---

[3]Satyendra Nath Bose (1894–1974), FRS, was an Indian physicist from the state of West Bengal, specializing in mathematical physics. He is best known for his work on quantum mechanics in the early 1920s, providing the foundation for Bose–Einstein statistics and the theory of the Bose–Einstein condensate. He is honoured as the namesake of the boson.

Although more than one Nobel Prize was awarded for research related to the concepts of the boson, Bose–Einstein statistics and Bose–Einstein condensate–the latest being the 2001 Nobel Prize in Physics, which was given for advancing the theory of Bose–Einstein condensates–Bose himself was never awarded the Nobel Prize. Among his other talents, Bose spoke several languages and could also play the Esraj, a musical instrument similar to a violin. [From Wikipedia]

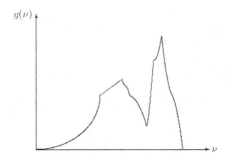

Fig. 2.5 Calculated phonon density of states for a sc structure.

is to determine the energy spectrum from the thermodynamic quantities. The specific heat of lattice vibrations can be expressed as

$$C_v(T) = rk \int_0^\infty \frac{(h\nu/kT)^2 e^{h\nu/kT}}{(e^{h\nu/kT} - 1)^2} g(\nu) d\nu \qquad (2.14)$$

where h and k represent the Planck constant and Boltzmann constant, respectively, T is the absolute temperature, r is the number of atoms per unit cell, $g(\nu)$ is phonon density of states which can be normalized to 1

$$\int_0^\infty g(\nu) d\nu = 1. \qquad (2.15)$$

The problem is to recover the $g(\nu)$ based on the experimentally measurable $C_v(T)$. This problem was proposed and approximately solved by Einstein in 1907 and Debye in 1912 using trial and error. Their contribution was crucial to the development of the concept of early quantum theory[Res85]; it has also been very important in the field of condensed matter physics. Einstein proposed a harmonic model with single frequency, which was a significant contribution to early quantum theory at the beginning of the 20th century[Ein07]. Soon after that, Debye suggested the continuous medium model for the low-temperature limit [Deb12]. After Blackman's calculation in 1935 (see Fig. 2.5), this problem was proposed again by Montroll[Mon42] in 1942 due to the importance of phonon density of states for thermodynamic properties of solids, lattice dynamics, electron-phonon interactions, and the optical-phonon spectrum. Lifshitz proposed the inverse specific heat problem independently in 1954[Lif54]. Montroll and Lifshitz[4] arrived at virtually identical solutions to this problem despite

---

[4]E. Montroll was an American mathematician, statistical physicist, and chemist. I.M. Lifshitz (1917–1982) was a Russian theoretical physicist.

the fact that the two worked in complete isolation as a result of World War II. It was repeated by Dai in 1990[Dai90]. Note that Lifshitz also obtained a formal solution by using Mellin transform in his work in 1954. In 1959, Weiss gave a general formula of the phonon density of states for low-frequency limit[Wei59]. Recently, Hughes, Frankel, and Ninham again used Mellin transform to obtain an integral representation of solution with the Weiss formula[Hug90]. Most of the works mentioned above focus on the integral representation of the exact solution with complex variables. As a result, these solutions are difficult to interpret in terms of intuition in physics. Also, different numerical methods to approximately determine the phonon density of states(DOS) from heat capacity data were proposed, for example, Chambers used high-order temperature derivatives to get rapid convergence at high temperature[Cha61], and Loram extracted the result from high-order moments of DOS [Lor86].

The present work introduces the Möbius inversion formula to obtain a concise and unified solution with real variable for this interesting inverse problem. This solution makes the discussion for various physical situations easier and much more convenient. Two general formulas for both high-frequency limit and low-frequency limit are given directly, and the Debye model and the Einstein model appear as two zero-order approximations. For convenience, we are going to introduce the previous work of Montroll first.

2.3.2 Montroll solution

In a crystal containing N atoms, the frequency distribution function $g(\nu)$ can defined as the number of canonical modes of vibration in the interval between ν and $\nu + d\nu$, i.e., $3Ng(\nu)d\nu$. The logarithm of the partition function per canonical mode is

$$\ln Z(T) = -\int_0^\infty g(\nu) \ln(1 - e^{-h\nu/kT}) d\nu. \tag{2.16}$$

Thus some physical quantities of interest can be given as integral equations related to $g(\nu)$. For example, the average energy of each canonical mode is

$$E(T) = kT^2 \frac{\partial \ln Z}{\partial T} = \int_0^\infty \frac{h\nu g(\nu) d\nu}{e^{h\nu/kT} - 1} \tag{2.17}$$

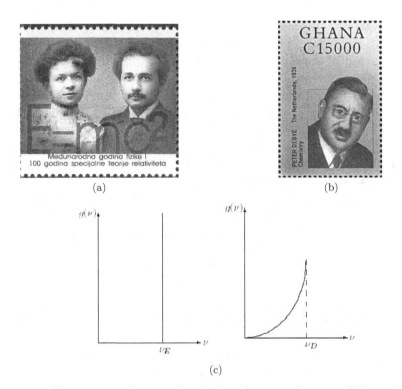

Fig. 2.6 (a) A. Einstein (1879–1955) and Mileva (1875–1948) in 1905, (b) Peter Joseph William Debye (1884–1966), a Dutch physicist and physical chemist, (c)Results from Einstein(1907) and Debye(1912).

According to the definition of heat capacity, we have

$$C_v(T) = (\frac{\partial E}{\partial T})_v = k \int_0^\infty (\frac{h\nu}{2kT})^2 \frac{g(\nu)d\nu}{\sinh^2(h\nu/2kT)} \qquad (2.18)$$

Introducing $\theta = h/kT$, then

$$\ln Z(T) = Z(\theta); \quad E(\theta) = \theta \, E(T)/h; \quad C(\theta) = C_v(T)/k \qquad (2.19)$$

Therefore, three integral equations are given

$$Z(\theta) = -\int_0^\infty g(\nu)\log(1 - e^{-\theta\nu})d\nu, \qquad (2.20)$$

$$E(\theta) = \int_0^\infty \frac{g(\nu)\theta\nu d\nu}{e^{\theta\nu} - 1}, \qquad (2.21)$$

Fig. 2.7 Elliott Montroll (1916–1983) and G.H. Weiss (1928–)

$$C(\theta) = \int_0^\infty \frac{(\theta\nu)^2 g(\nu) d\nu}{4\sinh^2(\theta\nu/2)} \qquad (2.22)$$

All of three can be put in one as

$$f(\theta) = \int_0^\infty g(\nu) K(\theta\nu) d\nu. \qquad (2.23)$$

Now we are going to solve it. Assuming that the functions $f(\theta)$ and $K(\theta\nu)$ are known. For conducting a Fourier transform, we change the variables at first

$$\theta = e^{-\eta} \text{ and } \nu = e^\alpha \qquad (2.24)$$

then

$$e^{-\eta} f(e^{-\eta}) = \int_{-\infty}^\infty g(e^\alpha) K(e^{\alpha-\eta}) e^{\alpha-\eta} d\alpha \qquad (2.25)$$

Multiplying both sides by $\exp\{-iu\eta\} d\eta/\sqrt{2\pi}$, and then we take the integral,

$$\frac{1}{\sqrt{2\pi}} \int_{-\infty}^\infty e^{-\eta} f(e^{-\eta}) e^{-iu\eta} d\eta$$

$$= \frac{1}{\sqrt{2\pi}} \int_{-\infty}^\infty g(e^\alpha) e^{-iu\alpha} d\alpha \int_{-\infty}^\infty e^{\alpha-\eta} e^{-iu(\eta-\alpha)} K(e^{\alpha-\eta}) d\eta$$

$$= \frac{1}{\sqrt{2\pi}} \int_{-\infty}^\infty g(e^\alpha) e^{-iu\alpha} d\alpha \int_{-\infty}^\infty e^{-\beta} e^{-iu(\beta)} K(e^{-\beta}) d\beta$$

where $\beta = \eta - \alpha$.
Define
$$I(u) = \int_{-\infty}^{\infty} e^{-\beta(iu+1)} K(e^{-\beta}) d\beta = \int_{0}^{\infty} x^{iu} K(x) dx \qquad (2.26)$$

Therefore,
$$\frac{1}{\sqrt{2\pi}} \int_{-\infty}^{\infty} e^{-\eta} f(e^{-\eta}) K(e^{\alpha-\eta}) e^{-iu\eta} d\eta = \frac{I(u)}{\sqrt{2\pi}} \int_{-\infty}^{\infty} g(e^{\alpha}) e^{-i\alpha u} d\alpha \qquad (2.27)$$

Carrying on Fourier transform, we have
$$g(e^{\alpha}) = \frac{1}{2\pi} \int_{-\infty}^{\infty} \frac{e^{iu\alpha}}{I(u)} \int_{-\infty}^{\infty} f(e^{-\eta}) e^{-\eta(1+iu)} d\eta.$$

Hence,
$$g(\nu) = \frac{1}{2\pi} \int_{-\infty}^{\infty} \frac{du}{I(u)} \int_{0}^{\infty} f(\theta)(\theta\nu)^{iu} d\theta \qquad (2.28)$$

The deduction for different $K(x)$ is as follows.
(1) $K(x) = \ln(1 - e^{-x})$
From
$$\ln(1 - e^{-x}) = -\sum_{n=1}^{\infty} \frac{e^{-nx}}{n}$$

it follows that
$$\int_{0}^{\infty} x^{iu} \ln(1 - e^{-x}) dx = -\int_{0}^{\infty} x^{iu} \sum_{n=1}^{\infty} \frac{e^{-nx}}{n} dx$$
$$= -\sum_{n=1}^{\infty} \int_{0}^{\infty} [nx]^{iu} \frac{e^{-nx}}{n^{2+iu}} d[nx] = -\sum_{n=1}^{\infty} \frac{1}{n^{2+iu}} \int_{0}^{\infty} y^{iu} e^{-y} dy$$
$$= -\zeta(2+iu)\Gamma(1+iu).$$

Therefore, we have

Theorem 2.2.
$$g(\nu) = \frac{1}{2\pi} \int_{-\infty}^{\infty} \frac{du}{\zeta(2+iu)\Gamma(1+iu)} \int_{0}^{\infty} Z(\theta)(\nu\theta)^{iu} d\theta \qquad (2.29)$$

(2) $K(x) = \dfrac{x}{(e^x - 1)}$

From

$$x/(e^x - 1) = \sum_{n=1}^{\infty} e^{-nx}$$

it follows that

$$\int_0^{\infty} x^{iu+1} \frac{1}{e^x - 1} dx = \sum_{n=1}^{\infty} \int_0^{\infty} [nx]^{iu+1} \frac{e^{-[nx]}}{n^{iu+2}} d[nx]$$
$$= \zeta(2 + iu)\Gamma(2 + iu)$$

Therefore we have

Theorem 2.3.

$$g(\nu) = \frac{1}{2\pi} \int_{-\infty}^{\infty} \frac{du}{\zeta(2+iu)\Gamma(2+iu)} \int_0^{\infty} E(\theta)(\nu\theta)^{iu} d\theta \qquad (2.30)$$

This solution is similar to that in the inverse blackbody radiation.

(3) $K(x) = \dfrac{x^2}{4\sinh^2 x/2}$

Similarly, we have

$$\int_0^{\infty} x^{iu+2} \frac{1}{4\sinh^2 x/2} dx = \zeta(2 + iu)\Gamma(3 + iu)$$

Therefore, we have

Theorem 2.4.

$$g(\nu) = \frac{1}{2\pi} \int_{-\infty}^{\infty} \frac{du}{\zeta(2+iu)\Gamma(3+iu)} \int_0^{\infty} C(\theta)(\nu\theta)^{iu} d\theta \qquad (2.31)$$

Montroll's formal solution indicates clearly that the density of states of phonon or the density of the vibration modes can be obtained by double integrals of heat capacity data. This formula involves complex variables; the complexity in the calculation can be imagined.

2.3.3 The Möbius formula on inverse heat capacity problem

Now let us introduce the new solution of inverse heat capacity problem by using Möbius series inversion formula. The new expression uses only real variables[Che90]. It is not only concise in form, but also reflects the unification of the Einstein approximation and the Debye approximation[Che98]. Similar to before, introducing a new variable, the coldness $u = h/kT$, then Eq. (2.18) becomes

$$C_v(\frac{h}{ku}) = rk \int_0^\infty \frac{(u\nu)^2 e^{u\nu}}{(e^{u\nu} - 1)^2} g(\nu) d\nu \qquad (2.32)$$

By using Taylor expansion, it follows that

$$C_v(\frac{h}{ku}) = rk \sum_{n=1}^\infty \int_0^\infty n(u\nu)^2 e^{-nu\nu} g(\nu) d\nu$$

$$= rku^2 \sum_{n=1}^\infty n \int_0^\infty \nu^2 e^{-nu\nu} g(\nu) d\nu$$

$$= rku \sum_{n=1}^\infty (nu) \mathcal{L}[\nu^2 g(\nu); \nu \to nu]$$

where $\mathcal{L}[\ ]$ represents the Laplace operator. Then from Möbius series inversion formula we have

$$u\mathcal{L}[\nu^2 g(\nu); \nu \to u] = \frac{1}{rk} \sum_{n=1}^\infty \mu(n) \frac{C_v(h/nku)}{nu} \qquad (2.33)$$

Finally, we have

Theorem 2.5.

$$g(\nu) = \frac{1}{rk\nu^2} \sum_{n=1}^\infty \mu(n) \mathcal{L}^{-1}[\frac{C_v(h/nku)}{nu^2}; u \to \nu]. \qquad (2.34)$$

Unlike Einstein's or Debye's approximation, Eq. (2.34) is an exact closed form solution to the inverse specific heat or the integral equation, Eq. (2.14). Like the Montroll solution, which includes double integrals with complex variables in integrand function, the inverse Laplace transform in the Möbius solution also implies integral with complex variables. But for some important cases in physics, the Möbius solution can give concise and clear results. Now let us show the rich and varied meaning of this formula.

Fig. 2.8 Pierre-Simon Laplace (1749–1827)

2.3.4 *General formula for the low temperature limit*

If we assume a standard low-temperature expansion of the specific heat in odd powers of T, we may write

$$C_v(T) = a_3 T^3 + a_5 T^5 + a_7 T^7 + \cdots \qquad T \to 0 \qquad (2.35)$$

or by using the new variable u

$$C_v(h/ku) = \sum_{n=2}^{\infty} a_{2n-1} u^{-(2n-1)} \qquad u \to \infty. \qquad (2.36)$$

Therefore,

$$\begin{aligned}
g(\nu) &= \frac{1}{rk\nu^2} \sum_{n=1}^{\infty} \mu(n) \sum_{m=2}^{\infty} a_{2m-1} \left(\frac{h}{nk}\right)^{(2m-1)} \mathcal{L}^{-1}\left[\frac{u^{-(2m-1)}}{nu^2}; u \to \nu\right] \\
&= \frac{1}{rk\nu^2} \sum_{m=2}^{\infty} \left[\sum_{n=1}^{\infty} \frac{\mu(n)}{n^{2m}}\right] a_{2m-1} \left(\frac{h}{k}\right)^{(2m-1)} \mathcal{L}^{-1}\left[\frac{1}{u^{(2m+1)}}; u \to \nu\right] \\
&= \frac{1}{rk\nu} \sum_{m=2} \left[\frac{1}{\zeta(2m)}\right] a_{2m-1} \left(\frac{h\nu}{k}\right)^{2m-1} \frac{1}{(2m)!}.
\end{aligned} \qquad (2.37)$$

Finally, by using

$$\zeta(2m) = (-1)^{m+1} \{(2\pi)^{2m}/[2(2m)!]\} B_{2m}$$

we have

Theorem 2.6.

$$g(\nu) = \frac{2}{rk\nu} \sum_{m=2}^{\infty} \frac{a_{2m-1}(h\nu/k)^{2m-1}}{(-1)^{(m+1)}(2\pi)^{2m} B_{2m}} \qquad (2.38)$$

This result is just the same as Weiss [Wei59]. It covers much more physics than Debye's solution. If only the first term on the right hand in Eq. (2.38) is taken, then

$$g(\nu) = \frac{2}{rk\nu} \frac{(-1)^{2+1} a_3 (h\nu/k)^3}{(2\pi)^4 B_4} = \frac{2}{rk\nu} \frac{(-1)^3 a_3 (h\nu/k)^3}{(2\pi)^4 (-1/30)}$$

or

Theorem 2.7.

$$g(\nu) = \left[\frac{15 a_3 h^3}{4\pi^4 k^4 r}\right] \nu^2 \qquad (2.39)$$

This is just the Debye approximation. From experimentally measurable a_3, the parabolic coefficient can be determined, then the Debye frequency or Debye temperature can be given by normalization condition Eq. (2.15).

2.3.5 Temperature dependence of Debye frequency

Corresponding to Eq. (2.39) there is only the first term in Eq. (2.35) as

$$C_v(T) = a_3 T^3$$

but the T^3 is only suitable for the limit case with $T \to 0$. For a certain temperature range of (0,T), one has to consider other terms in Eq. (2.39) which causes the temperature dependence of Debye frequency or the temperature dependence of Debye temperature.

Assume that the approximation of third degree is good enough as

$$C_v(T) = a_3 T^3 + a_5 T^5 + a_7 T^7$$

and define $\bar{a}_3(T)$ as the effective fitting coefficient of Debye approximation for temperature range $(0, T)$

$$\bar{a}_3(T) = \frac{1}{T} \int_0^T \frac{a_3 t^3 + a_5 t^5 + a_7 t^7}{t^3} dt = a_3 \left[1 + \frac{1}{3}\left(\frac{a_5}{a_3}\right) T^2 + \frac{1}{5}\left(\frac{a_7}{a_3}\right) T^4\right].$$

Obviously, $\bar{a}_3(T)$ is temperature dependent, thus a temperature dependent phonon spectrum is given by

$$g(\nu) = [\frac{15 \bar{a}_3(T) h^3}{4\pi^4 k^4 r}] \nu^2.$$

By using normalization condition, it is simply obtained as

$$\left(\frac{\nu_D(T)}{\nu_D(0)}\right)^3 = \frac{1}{1 + \frac{1}{3}\left(\frac{a_5}{a_3}\right) T^2 + \frac{1}{5}\left(\frac{a_7}{a_3}\right) T^4}$$

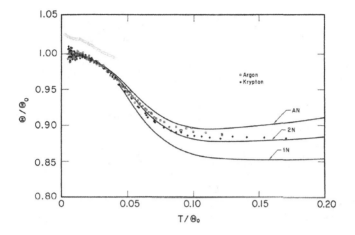

Fig. 2.9 Temperature dependence of Debye frequency

thus it is given by

Theorem 2.8.

$$\frac{\nu_D(T)}{\nu_D(0)} = \left[\frac{1}{1 + \frac{1}{3}(\frac{a_5}{a_3})T^2 + \frac{1}{5}(\frac{a_7}{a_3})T^4}\right]^{1/3}$$
$$\simeq 1 - \frac{1}{9}(\frac{a_5}{a_3})T^2 + \frac{1}{9}\left[\frac{2}{9}\left(\frac{a_5}{a_3}\right)^2 - \frac{3}{5}\left(\frac{a_7}{a_3}\right)\right]T^4.$$

This expression describes the temperature dependence of Debye frequency. As temperature range increases, the Debye frequency decreases as the term of T^2, then rises a little bit due to the term of T^4. The corresponding experiment data can be referred to Fig. 2.9[Fin69].

2.3.6 General formula for high temperature limit

Now let us show how to get a general formula for the high-frequency limit. If the temperature is high enough, we have

$$C_v(T) = a_0 - \frac{a_2}{T^2} + \frac{a_4}{T^4} - \cdots \qquad (2.40)$$

To replace T by coldness u, then

$$\frac{C(h/ku)}{u^2} = \frac{1}{u^2}[a_0 - \frac{a_2(ku)^2}{h^2} + \frac{a_4(ku)^4}{h^4} - \cdots] \qquad (2.41)$$

Therefore,
$$\frac{C(h/nku)}{nu^2} = \sum_{m=0}^{\infty}(-1)^m a_{2m}(k/h)^{2m}u^{2(m-1)}n^{2(m-1)} \tag{2.42}$$

Substituting this expression into Eq. (2.34), it is given by

$$g(\nu) = \frac{1}{rk\nu^2}\sum_{n=1}^{\infty}\mu(n)\mathcal{L}^{-1}[\frac{C(h/nku)}{nu^2}; u \to \nu]$$

$$= \frac{1}{rk\nu^2}\sum_{n=1}^{\infty}\mu(n)\mathcal{L}^{-1}[\frac{a_0}{nu^2}] - \frac{1}{rk\nu^2}\sum_{n=1}^{\infty}\mu(n)\mathcal{L}^{-1}[\frac{na_2k^2}{h^2}]$$

$$+ \frac{1}{rk\nu^2}\sum_{n=1}^{\infty}\mu(n)\mathcal{L}^{-1}[\frac{n^3u^2a_4k^4}{h^4}] - \cdots$$

$$= \frac{1}{rk\nu^2}\{[\sum_{n=1}^{\infty}\frac{\mu(n)}{n}]\mathcal{L}^{-1}[\frac{a_0}{u^2}] - [\sum_{n=1}^{\infty}n\mu(n)]\mathcal{L}^{-1}[\frac{a_2k^2}{h^2}]$$

$$+ [\sum_{n=1}^{\infty}n^3\mu(n)]\mathcal{L}^{-1}[\frac{a_4k^4u^2}{h^4}] - \cdots\}$$

$$= \frac{1}{rk\nu^2}\{\frac{1}{\zeta(1)}\mathcal{L}^{-1}[\frac{a_0}{u^2}] - \frac{1}{\zeta(-1)}\mathcal{L}^{-1}[\frac{a_2k^2}{h^2}] + \frac{1}{\zeta(-3)}\mathcal{L}^{-1}[\frac{a_4k^4u^2}{h^4}] - \cdots\}.$$

Notice that in the above deduction the following relations of generalized functions have been used including

$$\zeta(1) = \infty \tag{2.43}$$

with

$$\mathcal{L}^{-1}[u^m; u \to \nu] = \delta^{(m)}(\nu) \tag{2.44}$$

and

$$\frac{1}{\zeta(-s)} = \sum_{n=1}^{\infty}\mu(n)n^s \tag{2.45}$$

in which $s = 1, 2, 3, \ldots$. Note that the condition $s < 1$ may cause suspicion, but the unreasonable effectiveness of Eq. (2.44) can be understood as a result of a generalized function (see next subsection). Then the new general solution of inverse heat capacity for high temperature regime appears as

Theorem 2.9.

$$g(\nu) = \frac{1}{rk\nu^2}\sum_{n=1}^{\infty}\frac{a_{2n}}{\zeta(1-2n)}(\frac{k}{h})^{2n}\delta^{(2n-2)}(\nu) \tag{2.46}$$

Fig. 2.10 Niels Henrik Abel (1802–1829)

This is a general formula for high temperature limit by using Möbius series inversion with the Poisson–Abel principal value. Figure 2.10 shows a postage stamp with the image of Abel.

(1) Einstein single peak approximation

Einstein approximation would be the most important check for our general high-frequency formula. In this case, Eq. (2.46) can be rewritten as[5]

$$g(\nu) = \frac{1}{rk\nu^2}\{\frac{a_2}{\zeta(-1)}(\frac{k}{h})^2\delta(\nu) + \frac{a_4}{\zeta(-3)}(\frac{k}{h})^4\delta^{(2)}(\nu) + \cdots\}$$

$$= \frac{k}{rh^2\nu^2}\frac{a_2}{[-\zeta(-1)]}[\delta(\nu) + \frac{\nu_E^2}{2}\delta^{(2)}(\nu) + \cdots]$$

$$\simeq \frac{k}{rh^2\nu^2}\frac{a_2}{[-\zeta(-1)]}\delta(\nu - \nu_E)$$

or

Theorem 2.10.

$$g(\nu) \simeq \frac{12ka_2}{rh^2\nu_E^2}\delta(\nu - \nu_E) \qquad (2.47)$$

Thus the Einstein frequency ν_E can be obtained directly from experimental data a_2, a_4 as

Theorem 2.11.

$$\nu_E = \frac{k}{h}(\frac{2a_4[-\zeta(-1)]}{a_2\zeta(-3)})^{1/2} = \frac{k}{h}\sqrt{\frac{20a_4}{a_2}} \qquad (2.48)$$

This indicates that the Einstein frequency is dependent only on the first two expansion coefficients in series Eq. (2.46). In the solution Eq. (2.47), we have considered that $a_{2m} > 0$ and $\zeta(1-2m) = -B_{2m}/2m$, such as from

$$B_2 = 1/6, \ B_4 = -1/30, \ B_6 = 1/42, \ B_8 = -1/30, \ B_{10} = 5/66, \ldots$$

---

[5] $\delta(\nu)$ means Dirac δ function

to obtain

$$\zeta(-1) = -\frac{1}{12}, \ \zeta(-3) = \frac{1}{120}, \ \zeta(-5) = \frac{-1}{252}, \ \zeta(-7) = \frac{1}{240}, \ \zeta(-9) = \frac{-1}{132}, \cdots$$

In fact, Einstein's solution $\delta(\nu - \nu_E)$ in Eq. (2.41) corresponds to the heat capacity spectrum[Res85]

$$\frac{x^2 e^x}{(e^x - 1)^2} = 1 - \frac{x^2}{12} + \frac{x^4}{240} - \frac{x^6}{6080} + \cdots \quad (x = T_E/T) \qquad (2.49)$$

Substituting this expression into Eq. (2.47), we obtain

$$\nu_E = \frac{k}{h}\left[\frac{20T - E^4/240}{T_E^2/12}\right]^{1/2} = \frac{kT_E}{h} \qquad (2.50)$$

This is completely the same as that of Resnick[Res85]. All these "coincidences" indicate the success and effectiveness of the Möbius inversion.

(2) Two peak approximation

Suppose that there are two peaks located at ν_+ and ν_- with the form as two Dirac $\delta-$ function, then it is given by

$$\frac{\nu_+^2 + \nu_-^2}{2!} = 2\frac{a_4\zeta(-1)}{-a_2\zeta(-3)}\left(\frac{k}{h}\right)^2 \equiv A \qquad (2.51)$$

and

$$\frac{\nu_+^4 + \nu_-^4}{4!} = 2\frac{a_6\zeta(-1)}{-a_2\zeta(-5)}\left(\frac{k}{h}\right)^4 \equiv B \qquad (2.52)$$

Hence,

$$\nu_+^2 + \nu_-^2 = 2A \qquad (2.53)$$

$$2\nu_+^2 \nu_-^2 = 4A^2 - 24B \qquad (2.54)$$

Therefore,

$$\nu_\pm = \sqrt{A \pm \sqrt{12B - A^2}} \qquad (2.55)$$

A more complicated case with three or four peaks can be discussed in a similar way.

From the above, the quite interesting and difficult inverse specific heat problem is solved unexpectedly by the apparently obscure Möbius technique in this concise manner. What is more? This solution not only put the solutions of Einstein, Debye, and Weiss together, but also gave better solution than Einstein's approximation, and others. This cannot be considered as an isolated and occasional situation[Bur91]. The wide use of the number theory technique reflects the arrival of the digital information age, and the quantum age. In addition, a power series representation for the Riemann function has been extended to negative integer points in a Poisson–Abel process, which is very useful for physical applications. This might become a typical technique to deal with some inherently ill-posed inverse problems instead of some numerical methods such as the maximum entropy method and the Tikhonov[6].

(3) The temperature dependence of Einstein frequency

Fig. 2.11   Andrei Nikolaevich Tikhonov (1906–93)

Similar to the case at low temperature, if the concept of Einstein temperature is applied to certain range (T, ∞), then the temperature dependence of Einstein temperature at high temperature occurs. Obviously, Eq. (2.41) can be rewritten as

$$a_0 - C_v(h/ku) = (\frac{ku}{h})^2 [a_2 - \overline{a_4}(h/ku)]$$

where $\overline{a_4}(h/ku)$ is approximately equal to

$$\overline{a_4}(h/ku) = \frac{1}{u} \int_0^u [\frac{a_4(ku/h)^4 - a_6(ku/h)^6 + a_8(ku/h)^8}{(ku/h)^4}] du$$

---

[6] Andrei Nikolaevich Tikhonov or Andrey Nokolayevich Tychnoff (1906–93), was a Russian mathematician. He published his first paper on conditions for a topological space to be metrisable. Later, Tikhonov's work led from topology to functional analysis with his famous fixed point theorem for continuous maps from convex compact subsets of locally convex topological space in 1935. In the 1960s Tikhonov began to produce an important series of papers on ill-posed problems. He defined a class of regularisable ill-posed problems and introduced the concept of a regularising operator which was used in the solution of many problems in mathematical physics. [From Wikipedia]

$$= a_4[1 - \frac{a_6(ku/h)^2}{3a_4} + \frac{a_8(ku/h)^4}{5a_4}].$$

Therefore, the temperature dependence of Einstein frequency can be expressed as

$$\nu_E(T) = \frac{k}{h}\sqrt{\frac{20\overline{a_4}(T)}{a_2}} = \nu_E(\infty)\sqrt{1 - \frac{a_6}{3a_4T^2} + \frac{a_8}{5a_4T^4}}.$$

Hence, it is given by

Theorem 2.12.

$$\nu_E(T) \cong \nu_E(\infty)[1 - \frac{a_6}{6a_4T^2} - \frac{1}{72}\frac{a_6^2}{a_4^2T^4} + \frac{a_8}{10a_4T^4}].$$

This is the temperature dependence of Einstein temperature, which increases as the lower limit T of range (T, ∞) goes down.

2.3.7 *Some special relations between $\zeta(s)$ and $\mu(n)$*

It is well known that there is a relation between $\zeta(s)$ and Möbius as

$$\frac{1}{\zeta(s)} = \sum_{n=1}^{\infty} \mu(n)\frac{1}{n^s}, \quad s > 1 \tag{2.56}$$

Now we are going to see the case of $s < 1$. For doing so, we define two divergent series with the condition of $x \in (0,1)$.

$$p(x,s) = \sum_{n=1}^{\infty} \mu(n)n^s x^n \qquad x \in (0,1), \tag{2.57}$$

and

$$q(x,s) = \sum_{n=1}^{\infty} n^s x^n \qquad x \in (0,1). \tag{2.58}$$

In order to evaluate the generalized functions

$$\sum_{n=1}^{\infty} n^s \quad (s > 1)$$

and

$$\sum_{n=1}^{\infty} \mu(n)n^s \quad (s > 1)$$

the Poisson–Abel principal value or the limit of the Poisson–Abel generalized sum

$$\lim_{x \to 1^-} p(x, s) = \lim_{x \to 1^-} \sum_{n=1}^{\infty} \mu(n) n^s x^n \qquad x \in (0, 1) \qquad (2.59)$$

and

$$\lim_{x \to 1^-} q(x, s) = \lim_{x \to 1^-} \sum_{n=1}^{\infty} n^s x^n \qquad x \in (0, 1) \qquad (2.60)$$

is taken. From now on, all the divergent series will be considered as Poisson–Abel generalized sums without explanation (some denotations are ignored).
(1) The regular relation between $\zeta(s)$ and $\mu(n)$.
The Riemann ζ function is defined as

$$\zeta(s) = \sum_{n=1}^{\infty} \frac{1}{n^s} \qquad s > 1 \qquad (2.61)$$

The another relation between $\zeta(s)$ and $\mu(n)$ has been proven to be

$$\frac{1}{\zeta(s)} = \sum_{n=1}^{\infty} \frac{\mu(n)}{n^s} \qquad s > 1 \qquad (2.62)$$

The question is what happen when $s < 0$.
(2) A special relation between two divergent sums
The product of two generalized functions can be treated in a generalized way no matter $s > 0$ or $s < 0$. Therefore, we have

$$\lim_{0 < x \to 1} [\sum_{n=1}^{\infty} \mu(n) n^s x^n][\sum_{m=1}^{\infty} m^s x^m] = \lim_{0 < x \to 1} [\sum_{m,n=1}^{\infty} \mu(n)(mn)^s x^{m+n}]$$

$$= \sum_{k=1}^{\infty} [\lim_{0 < x \to 1} \sum_{n|k} \mu(n) x^{n+k/n}] k^s = 1 \qquad (2.63)$$

then it follows

Theorem 2.13.

$$\lim_{0 < x \to 1} [\sum_{n=1}^{\infty} \mu(n) n^s x^n] = \frac{1}{\lim_{0 < x \to 1} [\sum_{m=1}^{\infty} m^s x^m]} \qquad (2.64)$$

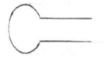

Fig. 2.12 The path of integral

(3) Other two pseudo-theorems on $\zeta(s)$

$$\lim_{0<x\to 1}\sum_{n=0}^{\infty}(n+1)^m x^{n+1} = -\frac{m!}{2\pi i}\int_{\infty}^{(0^+)}(-x)^{(m+1)}\frac{e^{-x}}{1-e^{-x}}d\,x$$

$$= -m![\frac{1}{2\pi i}\int_{\infty}^{(0^+)}(-x)^{(m+1)}\frac{e^{-x}}{1-e^{-x}}d\,x]$$

$$= (-m!)[(-1)^{(m+1)}\frac{B_{m+1}}{(m+1)!}] = \frac{(-1)^m B_{m+1}}{m+1} \qquad (2.65)$$

then

$$\lim_{0<x\to 1}\sum_{n=1}^{\infty}n^m x^n = \lim_{0<x\to 1}\sum_{n=0}^{\infty}(n+1)^m x^{n+1}$$

$$= (-1)^m \frac{B_{m+1}}{m+1} = \begin{cases} 0, & m=2k>0 \\ -\dfrac{B_{2k}}{2k}, & m=2k-1 \end{cases} \qquad (2.66)$$

In other words, the first pseudo-theorem is given by

Theorem 2.14.

$$\lim_{0<x\to 1}\sum_{n=1}^{\infty}n^m x^n = \begin{cases} 0, & m=2k>0 \\ \zeta(1-2k), & m=2k-1 \end{cases} \qquad (2.67)$$

Combining Eqs. (2.66) and (2.64), the second pseudo-theorem is obtained as

Theorem 2.15.

$$\sum_{n=1}^{\infty}\mu(n)n^{2k-1} = \frac{1}{\zeta(1-2k)}, \quad k=1,2,3,... \qquad (2.68)$$

(4) Detail from Eqs. (2.67) to (2.68)

Taking the integral representation of $\Gamma(s)$ for an arbitrary s as[Apo76]

$$\frac{1}{\Gamma(z)} = \frac{-1}{2\pi i}\int_\infty^{(0+)} e^{-t}(-t)^{-z}dt \tag{2.69}$$

in which the integral path is from infinity on the real axis, surrounding the origin counterclockwise once, and back to infinity on the real axis as in Fig. 2.12. Now let us consider the case of $z = m + 1$, where m is a positive integer. Introducing the variable t as

$$t = (n+1)x \tag{2.70}$$

we obtain that

$$\begin{aligned}
1 &= \frac{-\Gamma(z)}{2\pi i}\int_\infty^{(0+)} e^{-t}(-t)^{-z}dt \\
&= \frac{-\Gamma(m+1)}{2\pi i}\int_\infty^{(0+)} e^{-(n+1)x}[-x(n+1)]^{-(m+1)}(n+1)dx \\
&= \frac{-m!}{2\pi i}\int_\infty^{(0+)} e^{-(n+1)x}(n+1)^{-m}(-x)^{-(m+1)}dx.
\end{aligned}$$

Hence,

$$(n+1)^m = \frac{-m!}{2\pi i}\int_\infty^{(0+)} e^{-(n+1)x}(-x)^{-(m+1)}dx \tag{2.71}$$

Taking the summation, we have

$$\begin{aligned}
\lim_{0<y\to 1^-}\sum_{n=0}^\infty (n+1)^m y^{n+1} &= \frac{-m!}{2\pi i}\int_\infty^{(0+)}(-x)^{-(m+1)}\sum_{n=0}^\infty e^{-(n+1)x}dx \\
&= \frac{-m!}{2\pi i}\int_\infty^{(0+)}(-x)^{-(m+1)}e^{-x}\sum_{n=0}^\infty e^{-nx}dx \\
&= \frac{-m!}{2\pi i}\int_\infty^{(0+)}(-x)^{-(m+1)}\frac{e^{-x}}{1-e^{-x}}dx.
\end{aligned}$$

The integral in the last line has a pole of $(m+2)$th order. From this,

$$\begin{aligned}
I &= \frac{1}{2\pi i}\int_\infty^{(0+)}(-x)^{-(m+1)}\frac{e^{-x}}{1-e^{-x}}dx = \text{Res}_{x=0}[(-x)^{-(m+1)}\frac{e^{-x}}{1-e^{-x}}] \\
&= \frac{(-1)^{m+1}}{(m+1)!}\lim_{x\to 0}\frac{\partial^{(m+1)}}{\partial x^{(m+1)}}[\frac{x^{m+2}x^{-(m+1)}}{e^x - 1}] \\
&= \frac{(-1)^{m+1}}{(m+1)!}\lim_{x\to 0}\frac{\partial^{(m+1)}}{\partial x^{(m+1)}}[\frac{x}{e^x-1}].
\end{aligned}$$

Based on the generating function of Bernoulli B_n, thus

$$\frac{t}{e^t - 1} = \sum_{n=0}^\infty \frac{t^n}{n!}B_n \tag{2.72}$$

then

$$I = \frac{1}{2\pi i} \int_\infty^{(0+)} (-x)^{-(m+1)} \frac{e^{-x}}{1-e^{-x}} dx$$

$$= \frac{(-1)^{m+1}}{(m+1)!} \frac{\partial^{(m+1)}}{\partial x^{(m+1)}} [\sum_{n=0}^\infty \frac{x^n}{n!} B_n]|_{x=0}.$$

Finally

$$I = \{\frac{(-1)^{m+1}}{(m+1)!} \frac{\partial^{(m+1)}}{\partial x^{(m+1)}} [\frac{x^{m+1} B_{m+1}}{(m+1)!}]|_{x=0}\}$$

$$= \frac{(-1)^{m+1}}{(m+1)!} [\frac{(m+1)!}{(m+1)!} B_{m+1}] = (-1)^{m+1} [\frac{B_{m+1}}{(m+1)!}].$$

Therefore,

$$\lim_{0<x\to 1^-} \sum_{n=0}^\infty (n+1)^m x^{n+1} = -m! [(-1)^{m+1} \frac{B_{m+1}}{(m+1)!}]$$

$$= \frac{(-1)^m B_{m+1}}{m+1} \quad (2.73)$$

Therefore, for any natural number m,

$$\lim_{0<x\to 1^-} \sum_{n=1}^\infty n^m x^n = \frac{(-1)^m B_{m+1}}{m+1} =$$

$$= \begin{cases} 0, & m = 2k \\ \frac{-B_{2k}}{2k} = \zeta(1-2k) & m = 2k-1 \end{cases} \quad (2.74)$$

Note that $B_{2k-1} = 0$.

The above deduction can be considered as the origin of Eq. (2.39) in the subsection on the high temperature approximation. It is unexpected that from the Möbius inversion we can simply obtain some theorem on generalized function. The modified relation between $\zeta(s)$ and $\mu(n)$ combining with some process of principal values might be a way to treat some ill-posed equations, except the maximum entropy method and regularization method.

Feynman did point out that *"it is obvious that the mathematical reasonings which have been developed are of great power and use for physicists. On the other hand, sometimes the physicists' reasoning is useful for mathematicians[Fey67]."*

Up to now, only the Möbius inversion method can combine the different solutions from Einstein, Debye, and Weiss. But the inverse problem is never terminated. For example, the limits of the integral might be finite, the high frequency vibration might be non-harmonic, and the data of heat capacity might be incomplete with errors. In addition, for the complete crystalline materials, the phonon spectrum can be done by *ab initio* calculation now. But for some complex structures, this inverse problem for phonon density of states is still very important.

2.4 Some Inverse Problems Relative to Frequency Spectrum

2.4.1 *Inverse spontaneous magnetization problem*

The relation between temperature dependence of spontaneous magnetization $\Delta M(T)$ and density of states of spin wave $D(\nu)$ can be expressed as

$$\Delta M(T) = \mu_B g \int_0^\infty \frac{D(\nu) d\nu}{e^{h\nu/kT} - 1} \tag{2.75}$$

where h, μ_B, k, g represent Planck constant, Bohr magneton, Boltzmann constant and Lande g-factor respectively. From Eq. (2.75) there are two kinds of problem. The inverse problem is to determine $D(\nu)$ from measured $\Delta M(T)$. From mathematical view point the integral equation Eq. (2.75) is equivalent to Eq. (2.8), the deduction for solution Eq. (2.20) is also available in this case. But the Möbius method can provide the approximate solution with arbitrary order.

As before to introduce

$$u = \frac{h}{kT} \tag{2.76}$$

and the function transform

$$F(u) = \Delta M(\frac{h}{ku}) \tag{2.77}$$

from this,

$$F(u) = \mu_B g \int_0^\infty \frac{D(\nu) e^{-u\nu}}{1 - e^{-u\nu}} d\nu$$

$$= \mu_B g \sum_{n=1}^\infty \int_0^\infty D(\nu) e^{-nu\nu} d\nu$$

$$= \mu_B g \sum_{n=1}^\infty \mathcal{L}[D(\nu)\nu \mapsto nu] \tag{2.78}$$

where $\mathcal{L}$ is Laplace operator on the dual space of ν and nu. By using Möbius series inversion theorem we have

$$\mathcal{L}[D(\nu)\nu \mapsto u] = \frac{1}{\mu_B g}\sum_{n=1}^{\infty}\mu(n)F(nu) \qquad (2.79)$$

and

$$D(\nu) = \frac{1}{\mu_B g}\sum_{n=1}^{\infty}\mu(n)\mathcal{L}^{-1}[F(nu); u \mapsto \nu]$$

$$= \frac{1}{\mu_B g}\sum_{n=1}^{\infty}\mu(n)\mathcal{L}^{-1}[\Delta\, M(\frac{h}{knu}); u \mapsto \nu] \qquad (2.80)$$

where $\mathcal{L}^{-1}$ is the inverse Laplace operator from space u to space ν. In principle, once the experimental data of $\Delta\, M(T)$ is fitted in, one can obtain the density of states of spin wave immediately. For a set of experimental data, how to choose an appropriate function to fit the data is ill-posed, at least it must be Laplace invertible for using this procedure. This is important difference between numerical and analysis methods even though both methods are approximate.

In the condition of low temperature, the experiment on ferro-magnet shows[Gos61]

$$\Delta\, M(T) = a_1 T^{3/2} + a_2 T^{5/2} \qquad (2.81)$$

But according to the theory of spin waves there are two additional terms such that[Dai87]

$$\Delta\, M(T) = a_1 T^{3/2} + a_2 T^{5/2} + a_3 T^{7/2} + a_4 T^4 \qquad (2.82)$$

where the term of T^4 reflects interaction between spin waves. Therefore,

$$F(u) = b_1 u^{-3/2} + b_2 u^{-5/2} + b_3 u^{-7/2} + b_4 u^{-4}. \qquad (2.83)$$

From this,

$$D(\nu) = \frac{1}{\mu_B g}\sum_{n=1}^{\infty}\mu(n)\mathcal{L}^{-1}[\sum_{m=1}^{3} b_m(nu)^{-(m+1/2)} + b_4(nu)^{-4}; u \mapsto \nu]$$

$$= \frac{1}{\mu_B g}\sum_{n=1}^{\infty}\mu(n)[\sum_{m=1}^{3}\frac{b_m \nu^{m-1/2}}{n^{m+1/2}\Gamma(m+1/2)} + \frac{b_4 \nu^3}{n^4 \Gamma(4)}].$$

Hence,

$$D(\nu) = \frac{1}{\mu_B g}[\sum_{m=1}^{3}\frac{b_m \nu^{m-1/2}}{\zeta(m+1/2)\Gamma(m+1/2)} + \frac{b_4 \nu^3}{\zeta(4)\Gamma(4)}]. \qquad (2.84)$$

If there is only the first term $a_1 T^{3/2}$ or $b_1 u^{-3/2}$ in the expansion expression of $\Delta M(T)$ or $F(u)$, then we have

$$D(\nu) = \frac{b_1}{\mu_B g}\cdot\frac{\nu^{1/2}}{\zeta(3/2)\Gamma(3/2)} \propto \nu^{1/2}. \qquad (2.85)$$

This is in agreement with previous literature[Oma75].

2.4.2 Inverse transmissivity problem

The total power of radiation is related to the radiation frequency spectrum as

$$J(T) = \frac{h}{c^3} \int_0^\infty \frac{\nu^3 g(\nu) d\nu}{e^{h\nu/kT} - 1}. \tag{2.86}$$

The inverse transmissivity problem is to determine the radiation frequency spectrum $g(\nu)$ from the temperature dependence $J(T)$ of the total power. Substituting $D(\nu) = \nu^3 g(\nu)$ for Eq. (2.86), the structure in which is almost the same as Eq. (2.75). Thus we can solve it by Möbius inversion method. Now introducing $u - h/kT$ and $F(u) = J(h/ku)$, it is given by

$$D(\nu) = \frac{c^3}{h} \sum_{n=1}^\infty \mu(n) \mathcal{L}^{-1}[F(nu); u \mapsto \nu] \tag{2.87}$$

or

$$g(\nu) = \frac{c^3}{h\nu^3} \sum_{n=1}^\infty \mu(n) \mathcal{L}^{-1}[J(h/nku); u \mapsto \nu] \tag{2.88}$$

The result is obvious[Ji2006].

2.5 Summary

Chapter 2 mainly introduced the inverse boson system problems, such as the inverse blackbody radiation problem and inverse heat capacity problem.

(1) With respect to the study on blackbody radiation, there are three mutually inverse problems as shown in Fig. 2.13.

    (i) From the measured $W(\nu)$ at varying constant temperature $(a(T) = \delta(T - T_0))$ to determine the kernel;

    (ii) From $a(T)$ to calculate $W(\nu)$ based on the Boltzmann–Planck kernel in Eq. (2.1);

    (iii) From $W(\nu)$ to extract $a(T)$ based on the Boltzmann–Planck kernel in Eq. (2.1).

Planck solved problem (i) in 1900, providing such a resilient foundation to quantum theory that his valiant attempts over the subsequent 10 years to disprove his own results were futile[7]. Based on Planck's result,

---

[7]Planck tried to grasp the meaning of energy quanta, but to no avail. "My unavailing attempts to somehow reintegrate the action quantum into classical theory extended over

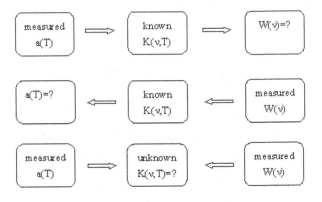

Fig. 2.13 Three kinds of typical inverse problems

the equation (2.1) was established, and the problem (ii) was proposed as a student's exercise in textbooks for generations. More recently in 1982, with respect to remote-sensing technique, Bojarsky addressed the problem (iii) as the inverse problem of problem (ii), that is, the so called inverse blackbody radiation problem discussed in this book. Bojarsky proposed an iterative procedure for solving this problem, and the Möbius inversion method gave a concise and general solution, which makes calculation and analysis simple.

On the other hand, from a mathematical point of view, the inverse blackbody radiation problem can be considered as a Fredholm integral equation of the first kind[Kir96] that is essentially ill-posed. In other words, solutions are very sensitive to errors of the input measured data, including data

---

several years and caused me much trouble." Even several years later, other physicists like Rayleigh, Jeans, and Lorentz set Planck's constant to zero in order to align with classical physics, but Planck knew well that this constant had a precise nonzero value. "I am unable to understand Jeans' stubbornness – he is an example of a theoretician as should never be existing, the same as Hegel was for philosophy. So much the worse for the facts, if they are wrong."

Max Born wrote about Planck: "He was by nature and by the tradition of his family conservative, averse to revolutionary novelties and skeptical towards speculations. But his belief in the imperative power of logical thinking based on facts was so strong that he did not hesitate to express a claim contradicting to all tradition, because he had convinced himself that no other resort was possible." [From Wikipedia]

incompleteness, and data inaccuracies. At first, Bojarski proposed an iterative method. Sun and Jaggard[Sun87] as well as Dou and Hodgson[Dou92a] used the Tikhonov regularization method to overcome these difficulties with some numerical examples. However, the unknown regularization parameter needed to be adjusted according to the area temperature distribution and to measurement errors. Dou and Hodgson[Dou92] applied the maximum entropy method to some simple examples, in their algorithm a potential function was introduced and the global minimum of the function was determined. Li[Li2005] formulated the inverse problem as an optimization problem, and the conjugate gradient method was used for its solution with a simple numerical example. In order to improve upon the solution, these different methods were applied, but one can not expect to find a complete one since the ill-posedness is inherent or permanent.

(2) Following and developing Planck's result, Einstein proposed an integral equation for lattice heat capacity, and provided a very natural δ-function solution to this inverse heat capacity problem. Both Einstein and Debye legitimized and developed the idea of quantum theory, so that their work has always occupied an important position in modern physics, though their solutions were narrowly focused, rough approximations. But as calculation and experiment unfolded, a corresponding general and accurate solution was required. Montroll was the first to attempt a general solution by using Fourier complex analysis, but this would be difficult to validate experimentally. A wave of efforts to improve upon his solution then appeared, among them Weiss's proposal of a generalized solution for low temperature approximation, which improved upon Debye's solution significantly. The Möbius inversion method, however, provided a general solution in real space, covering all of Einstein's, Debye's and Weiss's solutions as its special cases, and providing something new for the solution in high temperature regime. In particular, Chen and Rong[Che98] overcame the divergence of the inverse heat capacity problem by introducing a generalized theorem on Riemann ζ function. This yielded a very general formula for the solution as to the high temperature regime. Eventually, this method is equivalent to an asymptotic behavior control for cancelling the divergence in the theoretical model.

(3) Note that several closed-form solutions for inverse boson system problems based on the Möbius inversion method would be useful in testing purely numerical inversion results.

(4) The ill-posedness of an inverse problem not only implies the impossibility of obtaining a complete solution, but also suggests a variety of

possible ways to improve upon the existing solutions. Note that the definition of the inverse problems in mathematics is relatively narrow as compared to the much wider and uncertain definitions of inverse problems in physics. The reason is that physics is dependent not only on formal logical deduction, but also on intuition, analogy, concision, statistics, hypothesis emergent from new facts and experiment, and theory. For example, this book has omitted several interesting issues such as greybody radiation, nonharmonic vibrations in high frequency regime, the ill-posedness of Laplace inverse transform, and the λ- phase transition.

※※※※※※※※※※※※※※※※※※※※※※※※※※※※※※

A more interesting example is that of how it may be possible accurately to infer the frequency distribution of the vibrational states of a solid lattice, say $g(\nu)$, from measurements of specific heat at constant volume. Apart from numerical factors involving Planck's constant and Boltzmann's constant, the latter is simply the integral from zero to infinity of $\nu g(\nu)$ weighted by an appropriate Planck-Boltzmann factor allowing for the increased excitation of higher frequencies with increasing temperature. At low temperatures, it is usually feasible to represent the measured specific heat as a power series in the temperature T, beginning with a cubic term.

As if pulling a rabbit out of a hat, Chen relates $g(\nu)$ directly to the measured coefficients in the power series representing the specific heat. As a reminder that the theory of numbers lies at the basis of all this sleight of hand, values of Riemann's zeta function for integral multiples of 2 appear throughout, which means that they can be written as Bernoulli numbers. Chen's other example is that of inferring the temperature distribution of a composite black body from measurements of its power output, which he says is a problem of current interest in remote sensing.

Where will all this lead? The ideal, for Chen, would be that somebody should put a previously unsolved problem through the new Möbius mill. It will be interesting to see which problems first suggest themselves as candidates.

From John Maddox, Nature, Volume 344, 377, 29 March 1990

Chapter 3

Inverse Problems in Fermion Systems

3.1 The Arithmetic Functions of the Second Kind

In this chapter the so-called Möbius inversion method and Möbius function has been generalized to that accompanied with additive operations.

3.1.1 Definition of an arithmetic function of the second kind

Definition 3.1. An arithmetic function of the second kind is a function whose domain is the non-negative integers $\overline{Z^-}$, and whose range is a subset of complex numbers. Denote the whole set of arithmetic functions of the second kind as $\mathbb{A}_2$.

Definition 3.2. The binary operations between arithmetic functions f and g are defined by Dirichlet sum $\oplus$ as

$$\{f \oplus g\}(n) = \sum_{d=0}^{n} f(d)g(n-d) \qquad (3.1)$$

where the sum runs over all integers from 0 to n. For convenience, Definition 3.1 can be also written as

$$\{f \oplus g\}(n) = \sum_{a+b=n} f(a)g(b) \qquad (3.2)$$

where all the non-negative integer pair (a,b) sum to integers $a+b=n$. For example,

$$\{f \oplus g\}(0) = f(0)g(0),$$
$$\{f \oplus g\}(1) = f(0)g(1) + f(1)g(0),$$
$$\{f \oplus g\}(2) = f(0)g(2) + f(1)g(1) + f(2)g(0),$$
$$\{f \oplus g\}(3) = f(0)g(3) + f(1)g(2) + f(2)g(1) + f(3)g(0).$$

It is easy to prove the closeness of the Dirichlet sum, i.e.,

$$f, g \in \mathbb{A}_2 \Rightarrow f \oplus g \in \mathbb{A}_2 \qquad (3.3)$$

Similar to the set $\mathbb{A}$ associated with Dirichlet product $\otimes$ in Chapter 1, the set $\mathbb{A}_2$ with the corresponding binary operation $\oplus$ on it forms an algebraic system $\{\mathbb{A}_2, \oplus\}$, which is a semigroup. Now let us show the associativity of the arithmetic functions of the second kind:

$$f, g, h \in \mathbb{A}_2 \Rightarrow (f \oplus g) \oplus h(n) = f \oplus (g \oplus h)(n) \qquad (3.4)$$

where $n \in \overline{Z^-}$.

Proof. For any $n \in \overline{Z^-}$ we have

$$\{f \oplus g\} \oplus h(n) = \sum_{d+c=n} \{f \oplus g\}(d)h(c)$$

$$= \sum_{d+c=n} \sum_{a+b=d} f(a)g(b)h(c) = \sum_{a+b+c=n} f(a)g(b)h(c)$$

$$= \sum_{a+e=n} f(a) \sum_{b+c=e} g(b)h(c) = \sum_{a+e=n} f(a)\{g \oplus h\}(e)$$

$$= \{f \oplus (g \oplus h)\}(n). \qquad \square$$

The Dirichlet sum of the arithmetic functions of the second kind also satisfies commutativity. That is

$$f, g \in \mathbb{A}_2 \Rightarrow \{f \oplus g\}(n) = (g \oplus f)(n) \qquad (3.5)$$

where $n \in \overline{Z^-}$.

Now let us show some examples. $P_n(x) = \cos nx$, $P_n(x) = \sqrt{x^2 + n}$ and the derivative function $P_n(x) = Q^{(n)}(x)$ $(n = 0, 1, 2, \cdots)$ are arithmetic functions of the second kind, in particular, $A_n(x) = \sqrt{x^2 + n}$ and $P_n(x) = Q^{(n)}(x)$ are additive functions, since

$$P_m\{P_n(x)\} = \frac{d^m}{dx^m}\{\frac{d^n Q(x)}{dx^n}\} = \frac{d^{m+n} Q(x)}{dx^{m+n}} \qquad (3.6)$$

and

$$A_m A_n(x) = A_m(\sqrt{x^2 + n}) = \sqrt{x^2 + n + m} = A_{m+n}(x) \qquad (3.7)$$

But $P_n(x) = \cos nx$ is not an additive function.

3.1.2 Unit function in $\mathbb{A}_2$

For introducing the reversible function, we have to define the unit function at first.

Definition 3.3. For arbitrary $f \in \mathbb{A}_2$, there exists a function $\Delta_2 \in \mathbb{A}_2$ such that

$$f \oplus \Delta_2 = \Delta_2 \oplus f = f \tag{3.8}$$

Here Δ_2 is called unit function in $\mathbb{A}_2$. By using the induction it can be proven that in the algebraic system there exists only one function satisfying the condition of Δ_2, which is

$$\Delta_2(n) = \delta_{n,0} = \begin{cases} 1, & n = 0 \\ 0, & n \geq 1 \end{cases} \tag{3.9}$$

3.1.3 Inverse of an arithmetic function

Considering an arbitrary arithmetic function of the second kind as one of the elements of the semigroup $\mathbb{A}_2$, then how can we determine the existence of its inverse?

Definition 3.4. Assume that $f \in \mathbb{A}_2$, if there is a function $g \in \mathbb{A}_2$ such that $f \oplus g = \Delta_2$, then we call f and g mutually inverse, and denote g as f^{-1} and f as g^{-1}.

Obviously, there is not necessarily an inverse of arbitrary arithmetic function f. For example, there exists no inverse of the function f with $f(0) = 0$. It can be proven that the necessary and sufficient condition for the existence of inverse for f is that $f(0) \neq 0$.

Similarly, there is a constant unit function $\eta_2 \in \mathbb{A}_2$, $\eta_2(n) = 1$ for any $n \in \overline{Z^-}$. Also, η_2^{-1} satisfies

$$\sum_{d=0}^{n} \eta_2^{-1}(d) = \Delta_2(n) = \delta_{n,0} \tag{3.10}$$

For convenience, we define the Möbius function of the second kind as[1]

$$\mu_2(n) \equiv \eta_2^{-1}(n) = \begin{cases} 1, & n = 0 \\ -1, & n = 1 \\ 0, & n \geq 2 \end{cases} \tag{3.11}$$

---

[1] From the definition of Möbius function and Möbius inversion formula by G.C. Rota, this terminology should be allowed.

The corresponding inversion formula is called the Möbius inversion formula of the second kind.

Theorem 3.1.
$$g(n) = \sum_{d=0}^{n} f(d) \quad \Leftrightarrow \quad f(n) = \sum_{d=0}^{n} \mu_2(d) g(n-d) = g(n) - g(n-1) \quad (3.12)$$

3.2 Möbius Series Inversion Formula of the Second Kind

Now let us present some other Möbius series inversion formulas with their applications. The simplest one is

Theorem 3.2.
$$G(x) = \sum_{n=0}^{\infty} F(x+ny) \quad \Leftrightarrow \quad F(x) = \sum_{n=0}^{\infty} \mu_2(n) G(x+ny). \quad (3.13)$$

Notice that both $\mu_2(n)$ and $\eta_2(n)$ are not additive. Also, regarding the additive operation on the non-negative integers, there is only one "prime number 1"; this might be the reason that no one is interested in this kind of inversion formula. However, a mathematical "triviality" may not necessarily be useless in physics. Now let us introduce a generalized theorem.

Theorem 3.3.
$$G(x) = \sum_{n=0}^{\infty} r(n) F(x+ny) \quad \text{with} \quad r(0) \neq 0 \Leftrightarrow$$
$$F(x) = \sum_{n=0}^{\infty} r_\oplus^{-1}(n) G(x+ny) \quad (3.14)$$

where the inversion coefficient $r_\oplus^{-1}(n)$ satisfies
$$\sum_{n=0}^{k} r_\oplus^{-1}(n) r(k-n) = \delta_{k,0}. \quad (3.15)$$

Proof.
$$\sum_{n=0}^{\infty} r_\oplus^{-1}(n) G(x+ny) = \sum_{n=0}^{\infty} r_\oplus^{-1}(n) \sum_{m=0}^{\infty} r(m) F(x+ny+my)$$
$$= \sum_{k=0}^{\infty} \left\{ \sum_{n=0}^{k} r_\oplus^{-1}(n) r(k-n) \right\} F(x+ky)$$
$$= \sum_{k=0}^{\infty} \delta_{k,0} F(x+ky) = F(x). \quad \square$$

Similarly, we have[2]

Theorem 3.4.
$$G = r \oplus F \Leftrightarrow F = r_\oplus^{-1} \oplus G. \tag{3.16}$$

and

Theorem 3.5.
$$G(n,x) = \sum_{n=0}^{k} r(n) F(n,x) \Leftrightarrow F(n,x) = \sum_{n=0}^{k} r_\oplus^{-1}(n) G(n,x). \tag{3.17}$$

Now, consider the additivity of derivative operations,
$$\frac{d^n}{dx^n}\left(\frac{d^m}{dx^m}\right) = \frac{d^{m+n}}{dx^{m+n}} \tag{3.18}$$
This corresponds to another modified Möbius inversion theorem

Theorem 3.6.
$$G(x) = \sum_{n=0}^{\infty} r(n) F^{(n)}(x) \Leftrightarrow F(x) = \sum_{n=0}^{\infty} r_\oplus^{-1}(n) G^{(n)}(x) \tag{3.19}$$
where the coefficients satisfy $r(0) \neq 0$ and
$$\sum_{n=0}^{k} r_\oplus^{-1}(n) r(k-n) = \delta_{k,0}. \tag{3.20}$$

The proof is omitted.

3.3 Möbius Inversion and Fourier Deconvolution

The Fourier deconvolution
$$P(x) = \int_{-\infty}^{+\infty} Q(y) \Phi(y-x) dy \tag{3.21}$$
plays a very important role in many problems in science and technology. One of the problems is to extract $Q(x)$ from known $P(x)$ and $\Phi(x)$ based on Eq. (3.21). Denoting $\mathbb{F}[.]$ and $\mathbb{F}^{-1}[.]$ as the Fourier transform and inverse

---

[2] $r_\oplus^{-1} \oplus G = r_\oplus^{-1} \oplus [r \oplus F] = [r_\oplus^{-1} \oplus r] \oplus F = \Delta \oplus F = F.$

Fourier transform respectively, then the solution of the integral convolution Eq. (3.21) can be expressed as

$$Q(x) = \mathbb{F}^{-1}\left\{\frac{\mathbb{F}[P(x)]}{\mathbb{F}[\Phi(x)]}\right\} \qquad (3.22)$$

Now let us present a new method as follows. Taking Taylor's expansion of $Q(y)$ at x, we can rewrite Eq. (3.21) as[Xie95a; Xie95b; Che93]

$$P(x) = \int_{-\infty}^{\infty} \{\sum_{n=0}^{\infty} Q^{(n)}(x)\frac{(y-x)^n}{n!}\}\Phi(y-x)d(y-x)$$

$$= \sum_{n=0}^{+\infty} a(n)Q^{(n)}(x) \qquad (3.23)$$

where

$$a(n) = \frac{1}{n!}\int_{-\infty}^{+\infty} t^n \Phi(t)dt. \qquad (3.24)$$

By using the Möbius series inversion formula of the second kind, it follows that

$$Q(x) = \sum_{n=0}^{\infty} a^{-1}(n)P^{(n)}(x) \qquad (3.25)$$

the inversion coefficient $a^{-1}(n)$ can be determined by the following Dirichlet sum as

$$\sum_{n=0}^{k} a(n)a^{-1}(k-n) = \delta_{k,0}. \qquad (3.26)$$

The above inversion establishes a relation between two expansion bases in a functional space, one is $\{P^{(n)}(x), n = 0, 1, 2, ...\}$, another is $\{Q^{(n)}(x), n = 0, 1, 2, ...\}$. In general, the condition of this method is that Eq. (3.24) is held for arbitrary n. Obviously, this requires fast convergence of the kernel $\Phi(t)$ when $t \to \pm\infty$. For example,

$$\Phi(t) = O(e^{-|t|}) \quad \text{when} \quad t \to \pm\infty \qquad (3.27)$$

Regarding Eq. (3.26), there is an equivalent relation between $\{a(n)\}$ and $\{a^{-1}(n)\}$ as

Theorem 3.7.

$$\sum_{n=0}^{\infty} a(n)z^n = \frac{1}{\sum_{m=0}^{\infty} a^{-1}(m)z^m}. \qquad (3.28)$$

Proof.

$$\sum_{n=0}^{\infty} a(n)z^n \cdot \sum_{m=0}^{\infty} a^{-1}(m)z^m = \sum_{n=0}^{\infty} a(n) \cdot \sum_{m=0}^{\infty} a^{-1}(m)z^{m+n}$$

$$= \sum_{k=0}^{\infty} \left\{ \sum_{n=0}^{\infty} a(n) \cdot a^{-1}(k-n) \right\} z^k$$

$$= \sum_{k=0}^{\infty} \delta_{k,0} z^k = 1.$$

□

In brief, a new theorem on deconvolution is given by

Theorem 3.8.

$$P(x) = \int_{-\infty}^{+\infty} Q(y)\Phi(y-x)dy \Rightarrow Q(x) = \sum_{n=0}^{\infty} a^{-1}(n)P^{(n)}(x)$$

where $a^{-1}(n)$ is the inverse of $a(n)$ in Eq. (3.24). In fact, the solution, Eq. (3.25), is easy to expressed as different order approximations, which are convenient for analysis and evaluation.

3.4 Solution of Fermi Integral Equation

3.4.1 Fermi integral equation

An integral equation in which the kernel can be expressed in terms of Fermi distribution is called a Fermi integral equation For example, in the following

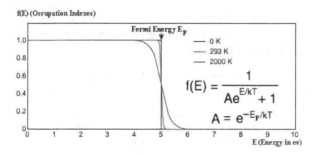

Fermi-Dirac distribution for several temperatures

Fig. 3.1 Fermi–Dirac distribution

Fig. 3.2 (a) Enrico Fermi (1901–1954), (b) Paul Dirac (1902–1984)

equation

$$P(x) = \int_{-\infty}^{\infty} \frac{Q(y)}{1+e^{y-x}} dy \qquad (3.29)$$

where $P(x)$ represents the total number of electrons, and $Q(y)$ the density of states. In fact, the inverse dielectric spectrum problem, inverse absorption problem and so on can be attributed to a similar integral equation.[3]

Now we are going to solve Eq. (3.29) for $Q(y)$. Based on Eqs. (3.23) and (3.24) we can simply obtain

$$P(x) = \sum_{n=0}^{\infty} a(n) Q^{(n)}(x)$$

where

$$a(n) = \frac{1}{n!} \int_{-\infty}^{\infty} \frac{t^n}{1+e^t} dt.$$

However, this integral equation is a little too complicated to evaluate and analyze since $1/(1+e^t)$ is neither an even function, nor an odd one. Thus we change the way as follows.

---

[3] Enrico Fermi (1901–1954) was an Italian physicist most noted for his work on the development of the first nuclear reactor, and for his contributions to the development of quantum theory, nuclear and particle physics, and statistical mechanics. He is acknowledged as a unique physicist who was highly accomplished in both theory and experiment. Fermium, a synthetic element created in 1952, the Fermi National Accelerator Lab, and a type of particles called fermions are named after him.

Paul Adrien Maurice Dirac was a British theoretical physicist. Dirac made fundamental contributions to the early development of both quantum mechanics and quantum electrodynamics. He held the Lucasian Chair of Mathematics at the University of Cambridge and spent the last 14 years of his life at Florida State University. [From Wikipedia]

Take the first derivative of both sides of Eq. (3.29) with respect to x, we have

$$P^{(1)}(x) = \int_{-\infty}^{\infty} \frac{e^{y-x}Q(y)}{[1+e^{y-x}]^2} dy. \tag{3.30}$$

Rewrite Eq. (3.30) as

$$\mathbb{P}(x) = P^{(1)}(x) = \sum_{n=0}^{\infty} \tilde{a}(n) Q^{(n)}(x) \tag{3.31}$$

where $\tilde{a}(n)$ is defined as

$$\tilde{a}(n) = \frac{1}{n!} \int_{-\infty}^{+\infty} t^n \tilde{\Phi}(t) dt = \frac{1}{n!} \int_{-\infty}^{+\infty} \frac{t^n e^t}{(1+e^t)^2} dt. \tag{3.32}$$

Now the solution of Eq. (3.29) becomes

$$Q(x) = \sum_{n=0}^{\infty} \tilde{a}^{-1}(n) \mathbb{P}^{(n)}(x). \tag{3.33}$$

Obviously, both $e^t/(1+e^t)^2$ and $t^{2m}e^t/(1+e^t)^2$ are even while $t^{2m-1}e^t/(1+e^t)^2$ is odd. This makes it easy to evaluate the integral in Eq. (3.32). From the odd–even property, it follows that

$$\tilde{a}(2m-1) = 0 \tag{3.34}$$

and

$$\tilde{a}(2m) = \frac{2}{(2m)!} \int_0^{\infty} t^{2m} \left\{ \sum_{k=1}^{\infty} (-1)^{k+1} k e^{-kt} \right\} dt$$

$$= 2 \sum_{k=1}^{\infty} \left\{ (-1)^{k+1} k \left[\frac{1}{(2m)!} \int_0^{\infty} t^{2m} e^{-kt} dt \right] \right\}$$

$$= 2 \sum_{k=1}^{\infty} \left\{ (-1)^{k+1} k \left[\frac{1}{(2m)!} (2m)! k^{-2m-1} \right] \right\}.$$

Hence,

$$\tilde{a}(2m) = 2 \sum_{k=1}^{\infty} \frac{(-1)^{k+1}}{k^{2m}} = 2(1-2^{1-2m})\zeta(2m) \tag{3.35}$$

Further more, we consider

$$\zeta(2m) = \frac{(-1)^{m+1} 2^{2m-1} \pi^{2m}}{(2m)!} B_{2m} \tag{3.36}$$

then

$$\tilde{a}(2m) = 2(1 - 2^{1-2m})\zeta(2m)$$
$$= \frac{(-1)^{m+1}(2^{2m} - 2)\pi^{2m}}{(2m)!} B_{2m} \quad (3.37)$$

Comparing with the Taylor's expansion of cosecant function[Kno28], we have

$$\frac{t}{\sin t} = \sum_{m=0}^{\infty} \frac{(-1)^{m+1}(2^{2m} - 2)\pi^{2m}}{(2m)!} B_{2m} t^{2m} \quad (3.38)$$

Therefore,

$$\sum_{n=0}^{\infty} a^{-1}(n) t^n = \frac{\sin t}{t} \quad (3.39)$$

and

$$\tilde{a}^{-1}(n) = \begin{cases} 0, & \text{if } n = 2m+1 \\ (-1)^m \pi^{2m}/(2m+1)! & \text{if } n = 2m \end{cases} \quad (3.40)$$

Now let us show another way to prove Eq. (3.40). According to Eq. (3.24),

$$\sum_{n=0}^{\infty} \tilde{a}(n) z^n = \sum_{n=0}^{\infty} \frac{1}{n!} \int_{-\infty}^{\infty} \frac{(zt)^n e^t}{(1+e^t)^2} dt \int_{0}^{\infty} \frac{t^z}{(1+t)^2} dt$$
$$= B(1-z, 1+z) = \frac{\Gamma(1-z)\Gamma(1+z)}{\Gamma(2)} = \frac{z\Gamma(1-z)\Gamma(z)}{\Gamma(2)}$$

hence

$$\sum_{n=0}^{\infty} \tilde{a}(n) z^n = \frac{\pi z}{\sin(\pi z)}$$

where $\Gamma(z)$ and $B(p,q)$ are defined as

$$\Gamma(z) = \int_0^{\infty} e^{-t} t^{z-1} dt \quad \text{and} \quad B(p,q) = 2 \int_0^{\pi/2} \sin^{2p-1}\theta \cos^{2q-1}\theta d\theta$$

respectively. It is easy to expand $\sin(\pi z)/\pi z$ to a power series of z. Then based on Eq. (3.28), it follows that

$$\sum_{n=0}^{\infty} \tilde{a}^{-1}(n) z^n = \frac{\sin(\pi z)}{\pi z} = \sum_{m=0}^{\infty} \frac{(-1)^m \pi^{2m}}{(2m+1)!} z^{2m}.$$

Thus the inversion coefficient $a^{-1}(n)$ is obtained as in Eq. (3.40).

Based on Eq. (3.33), the inversion formula for a Fermi system is given by

$$Q(x) = \frac{1}{\pi}\sum_{n=0}^{\infty}(-1)^n \frac{\pi^{2n+1}}{(2n+1)!}\frac{\partial^{2n+1}}{\partial x^{2n+1}}P(x) \tag{3.41}$$

Now let us apply the result to a typical inverse fermion system problem in semiconductor physics, namely

$$n(E_F, T) = \int_{-\infty}^{\infty} dE \, \frac{g(E,T)}{1+e^{(E-E_F)/kT}} \tag{3.42}$$

By using Eq. (3.40), we have

$$g(E_F, T) = \sum_{m=0}^{\infty}(-1)^m \frac{(\pi kT)^{2m}}{(2m+1)!}\frac{\partial^{2m+1}}{\partial E_F^{2m+1}}n(E_F, T) \tag{3.43}$$

In other words, one can determine the density of states of a fermion system based on the measurable carrier density and the Fermi level. In fact, the first three approximate solutions can be expressed as

$$g_0(E,T) = \frac{\partial n(E,T)}{\partial E}, \tag{3.44}$$

$$g_1(E,T) = \frac{\partial n(E,T)}{\partial E} - \frac{\pi^2}{6}(kT)^2\frac{\partial^3 n(E,T)}{\partial E^3}, \tag{3.45}$$

and

$$g_2(E,T) = \frac{\partial n(E,T)}{\partial E} - \frac{\pi^2}{6}(kT)^2\frac{\partial^3 n(E,T)}{\partial E^3} + \frac{\pi^4}{120}(kT)^4\frac{\partial^5 n(E,T)}{\partial E^5} \tag{3.46}$$

respectively. In fact, Eq. (3.33) may be written as

$$Q(x) = \frac{1}{\pi}\sin\left\{\left(\pi\frac{\partial}{\partial x}\right)P(x)\right\}$$

$$= \frac{1}{\pi}\Im[e^{i\pi\frac{\partial}{\partial x}}P(x)]$$

$$= \frac{1}{2i\pi}[P(x+i\pi) - P(x-i\pi)] \tag{3.47}$$

where $\Im$ represents the imaginary part. Later we will see that the solutions Eqs. (3.44) and (3.38), are also useful for the relaxation time spectrum[Fro58; Lig91], the distribution of Langmuir kernel[Sip50; Lan76; Cer80], and so on. Note that in Landman's work, the Wiener–Hopf method was applied. Also, it is interesting to notice that the multiplicative Möbius inversion reflects the inverse boson system problem, and the Möbius inversion of the second kind reflects the inverse fermion system problem. Note that we may not have complete data on $a(n)$ in practice, thus the $\{a^{-1}(n)\}$ can only be obtained partially.

3.4.2 Relaxation-time spectra

In general, the relaxation-time distribution $Y(\tau)$ and the measurable properties, the components of the complex permittivity $\epsilon(\omega) = \epsilon'(\omega) + \epsilon''(\omega)$ at frequency ω, can be expressed by

$$\int_0^\infty \frac{Y(\tau) d\tau}{1 + (\omega\tau)^2} \equiv \frac{\epsilon'(\omega) + \epsilon'_\infty}{\epsilon'_0 + \epsilon'_\infty} \equiv Z(\omega) \qquad (3.48)$$

where ϵ'_0 is the low-frequency (in the limit, static) permittivity of the material, and ϵ''_0 the high-frequency (optics) limit.

The traditional way of solving the above integral equation is to construct an expression of $Y(\tau)$ from general arguments with some parameters determined by experiments based on Eq. (3.48)[Fro58]. Thus the form of the unknown function has to be decided before doing the calculation. A new technique to calculate the spectra $Y(\tau)$ has been proposed by Ligachev and Falikov[Lig91] in 1991, but the method is related to both Mellin transform and the modified Bessel function of the third kind, and so on. Also, their fitting function is restricted[Lig91]. By using Eq. (3.47), we show a simple and general solution for this problem.

Denoting that

$$\tau^2 = e^x \quad and \quad \omega^2 = e^{-y} \qquad (3.49)$$

then

$$2\tau d\tau = e^x dx. \qquad (3.50)$$

Substituting Eqs. (3.49) and (3.50) into Eq. (3.48), it follows that

$$Z(e^{-y/2}) = \frac{1}{2} \int_{-\infty}^\infty \frac{Y(e^{x/2}) e^{x/2} dx}{1 + e^{x-y}}. \qquad (3.51)$$

Therefore,

$$\frac{1}{2} Y(e^{y/2}) e^{y/2} = \frac{1}{\pi} \sum_{m=0}^\infty \frac{(-1)^m \pi^{2m+1}}{(2m+1)!} \frac{\partial^{2m+1}}{\partial y^{2m+1}} Z(e^{-y/2})$$

$$= \frac{1}{\pi} \Im[\exp\left(i\pi \frac{\partial}{\partial y}\right) Z(e^{-y/2})]. \qquad (3.52)$$

The above expression is a general closed-form solution for Eq. (3.52). Now let us check the result of Ligachev and Filikov for the case of

$$Z(\omega) = \sum_{j=1}^N a_{kj} \exp\left[-\frac{b_{kj}}{2}\left(\frac{\omega}{\omega_j} + \frac{\omega_j}{\omega}\right)\right]. \qquad (3.53)$$

Let us only consider one term in Eq. (3.53), i.e.,

$$Z(e^{-y/2}) = \exp\left\{b\left[\frac{\omega}{\omega_0} + \frac{\omega_0}{\omega}\right]\right\}$$
$$= \exp\left\{-b\left(e^{(y-y_0)/2} + e^{-(y-y_0)/2}\right)\right\} \quad (3.54)$$

Based on Eq. (3.52), it follows that

$$Y(\tau) = \frac{2}{\pi\tau}\Im\left\{e^{i\pi(\partial/\partial y)}\exp\left[b(e^{(y-y_0)/2} + e^{-(y-y_0)/2})\right]\right\}\Big|_{y=x}$$
$$= \frac{2}{\pi\tau}\Im\left\{\exp[ib(e^{(y+i\pi-y_0)/2} + e^{-(y+i\pi-y_0)/2})]\right\}\Big|_{y=x}$$
$$= \frac{2}{\pi\tau}\Im\left\{\exp[ib(e^{(y-y_0)/2} - e^{-(y-y_0)/2})]\right\}\Big|_{y=x}$$
$$= \frac{2}{\pi\tau}\sin\left\{[b(e^{(y-y_0)/2} - e^{-(y-y_0)/2})]\right\}\Big|_{y=x}$$
$$= \frac{2}{\pi\tau}\sin\left\{[b(\frac{1}{\omega_0\tau} - \omega_0\tau)]\right\}\Big|_{y=x}. \quad (3.55)$$

This is just the same as that of Ligachev and Filikov, but the present deduction is much simpler and use only elementary operations.

3.4.3 Adsorption integral equation with a Langmuir kernel

The concept of adsorption on heterogeneous substrates can be traced to the pioneering work of Langmuir who proposed and expression for the total isotherm[Lan76; Bus2001]. Given the experimentally determined total

Fig. 3.3 Irving Langmuir (1881–1957)

isotherm Θ_t and a theoretical local therm Θ_L, it is necessary to evaluate the distribution function $\rho(\varepsilon, T)$, which satisfies

$$\Theta_t(P,T) = \int_0^\infty \Theta_L(P,T;\varepsilon)\rho(\varepsilon,T)d\varepsilon \qquad (3.56)$$

and

$$\int_0^\infty \rho(\varepsilon,T)d\varepsilon = 1 \qquad (3.57)$$

where $\varepsilon(\geq 0)$ is the absorption energy, P and T are pressure and temperature respectively. In general, it is assumed that the local isotherm Θ_L is the well known Langmuir isotherm

$$\Theta_L(P,T;\varepsilon) = [1 + P^{-1}a(T)\exp(-\varepsilon/RT)]^{-1} \qquad (3.58)$$

where the meaning of $a(T)$ can be explained from statistic derivation of Langmuir isotherm. Now the integral equation Eq. (3.56) becomes

$$\Theta_t(P,T) = \int_0^\infty \frac{\rho(\varepsilon,T)}{[1 + P^{-1}a(T)\exp(-\varepsilon/RT)]}d\varepsilon. \qquad (3.59)$$

Let

$$\rho(\varepsilon,T) \approx \rho(\varepsilon),\ P^{-1}a(T) = e^y[P = a(T)e^{-y}],\ \text{and}\ \ x = \varepsilon/RT \qquad (3.60)$$

then

$$\Theta_t\left(a(T)e^{-y}, T\right) = RT \int_0^\infty \frac{\rho(RTx)}{[1+e^{y-x}]}dx. \qquad (3.61)$$

Hence,

$$\rho(\varepsilon) = \rho(RTx) = \frac{-1}{\pi RT}\Im[e^{i\pi(\partial/\partial y)}\Theta_t(y,T)]|_{y=x}. \qquad (3.62)$$

Note that unlike $[1+e^{x-y}]^{-1}$ in Eq. (3.51), we are now facing $[1+e^{y-x}]^{-1}$ in Eq. (3.61). This is the reason why a minus sign appears in Eq. (3.62). This formulation can be also used to solve the common inverse Freundlich isotherm problem and inverse Dubinin–Radushkevich isotherm problem.

Fig. 3.4 Norbert Wiener (1894–1964): (a) in Tufts College at the age of eleven, and (b) in his lecture.

3.4.4 Generalized Freundlich isotherm

The corresponding empirical law can be expressed as

$$\Theta_t(P,T) = [1 + P^{-1}a(T)]^{-c} \quad 0 < c < 1. \tag{3.63}$$

From Eqs. (3.50) and (3.52), as well as

$$\frac{1}{(1+e^y)^c} = \frac{e^{-cy}}{(1+e^{-y})^c} \tag{3.64}$$

then

$$\rho(\varepsilon) = \rho(RTx) = \frac{-1}{2\pi i RT} \left\{ [e^{i\pi(\partial/\partial y)} - e^{-i\pi(\partial/\partial y)}] \frac{e^{-cy}}{[1+e^{-y}]^c} \right\}|_{y=x}$$

$$= \frac{-1}{2\pi i RT} \left\{ \frac{e^{-c(y+i\pi)}}{[1+e^{-y-i\pi}]^c} - \frac{e^{-c(y-i\pi)}}{[1+e^{-y+i\pi}]^c} \right\}|_{y=x}$$

$$= \frac{e^{-cy}}{2\pi RT} \frac{e^{ic\pi} - e^{-ic\pi}}{(1-e^{-y})^c} = \frac{\sin \pi c}{\pi RT[\exp(\varepsilon/RT) - 1]^c}. \tag{3.65}$$

This result is completely in agreement with that of Landmann([Lan76]) and Sips[Sip48; Sip50], but the present method is much more concise[Lan76]. Note that the transform Eq. (3.64) is necessary, since $(1 - e^y)^c$ is meaningless for $y > 0$.

3.4.5 Dubinin–Radushkevich isotherm

Dubinin–Radushkevich isotherm can be expressed as

$$\Theta_t(P,T) = \exp\{-B[RT\ln(P_0/P)]^2\} \tag{3.66}$$

Fig. 3.5 (a) Eberhard Hopf (1902–1983). (b) Uzi Landman (1944–)

where P_0 is the saturated vapor pressure of the absorbed gas at ambient temperature AT, and B is a constant.

$$A \equiv B(RT)^2 \quad \text{and} \quad C \equiv \ln[P_0/a(T)] \qquad (3.67)$$

then

$$\Theta_t\,(P,T) = \exp\{-A(y-C)^2\} \qquad (3.68)$$

Hence,

$$\begin{aligned}
\rho(\varepsilon) = \rho(RTx) &= \frac{-1}{\pi RT}\Im\left\{e^{i\pi(\partial/\partial y)}[e^{-A(y-C)^2}\right\}|_{y=x} \\
&= \frac{-1}{\pi RT}\left[e^{-A\{(y-C)-i\pi\}^2}\right]|_{y=x} \\
&= \frac{-1}{\pi RT}\left[e^{-A(y-C)^2+A\pi^2+2i\pi A(y-C)}\right]|_{y=x} \\
&= \frac{-1}{\pi RT}\left[e^{-A(y-C)^2+A\pi^2}\right]\sin 2\pi A(y-C)|_{y=\varepsilon/RT}.
\end{aligned}$$

Finally,

$$\rho(\varepsilon) = \frac{\exp\{-B\left[\varepsilon^2 - (\pi RT)^2\right]\}}{\pi RT}\sin(2\pi BRT\varepsilon) \qquad (3.69)$$

This is the same as the result by using the Stieltjes transform[Lan76].

3.4.6 Kernel expression by $\delta-$ function

This method is based on the discovery of a special operator, by which the integral kernel changes into δ function[Che93]. From the elementary theory of generalized function it is given by[Bog59; Sch61]

$$\lim_{\epsilon \to 0^+} \frac{1}{x \pm i\epsilon} = \frac{1}{x \pm i0^+} = P[\frac{1}{x}] \mp i\pi\delta(x) \qquad (3.70)$$

where P represents the Cauchy principal. Therefore,

$$\delta(x-y) = \frac{1}{2\pi i}\left[\frac{1}{x-y-i0^+} - \frac{1}{x-y+i0^+}\right]$$

$$= \frac{1}{2\pi i}\left[\frac{1}{1-e^{x-y+i0^+}} - \frac{1}{1-e^{x-y-i0^+}}\right] \qquad (3.71)$$

Introducing the translational operator

$$e^{i(\pi-0^+)\frac{\partial}{\partial y}} \qquad (3.72)$$

and

$$e^{-i(\pi-0^+)\frac{\partial}{\partial y}} \qquad (3.73)$$

they represent the translations of $\pm\pi$ respectively along the imaginary axis in the complex plane. Therefore,

$$\frac{1}{2\pi i}\left\{e^{i(\pi-0^+)\frac{\partial}{\partial y}} - e^{-i(\pi-0^+)\frac{\partial}{\partial y}}\right\}\frac{1}{1+e^{x-y}}$$

$$= \frac{1}{2\pi i}\left\{\frac{1}{1-e^{x-[y+i(\pi-0^+)]}} - \frac{1}{1-e^{x-[y-i(\pi-0^+)]}}\right\}$$

$$= \frac{1}{2\pi i}\left\{\frac{1}{1-e^{x-y+i0^+}} - \frac{1}{1-e^{x-y-i0^+}}\right\}$$

$$= \delta(x-y) \qquad (3.74)$$

Hence,

$$\delta(x-y) = \frac{1}{2\pi i}\left[e^{-i\pi(\partial/\partial y)} - e^{i\pi(\partial/\partial y)}\right]\frac{1}{1+e^{x-y}} \qquad (3.75)$$

or

$$\delta(x-y) = \frac{1}{\pi}\Im\{\exp[i\pi\frac{\partial}{\partial y}]\frac{1}{1+e^{x-y}}\} \qquad (3.76)$$

or

$$\delta(x-y) = \frac{1}{\pi}\sum_{m=0}^{\infty}\frac{(-1)^m \pi^{2m+1}}{(2m+1)!}\frac{\partial^{2m+1}}{\partial y^{2m+1}}\frac{1}{1+e^{x-y}} \qquad (3.77)$$

Since we have got the expression of Dirac-δ function by using the fermi distribution, the solution of a fermi integral equation can be obtained easily.

It is well known near by fermi surface, the DOS-temperature dependence $g(E_f, T)$ can be expressed as in Eq. (3.40), that is

$$g(E_f, T) = \sum_{m=0}^{\infty} \frac{(-1)^m (\pi k T)^{2m}}{(2m+1)!} \frac{\partial^{2m+1}}{\partial E_f^{2m+1}} n(E_f, T)$$

where $n(E_f, T)$ is the carrier density. This is just the solution of the fermi integral equation.

3.5 Möbius and Biorthogonality

In Chapter 1 the Chebyshev method was mentioned, based on which the relation between biothogonality and Möbius inversion will be presented in this section.

3.5.1 Chebyshev formulation

In 1851, Chebyshev did express a cosine function as a superposition of triangular waves. He started with the Fourier expansion of an even triangular wave $H_{tri}(t)$ having a period of 2π[Che1851]. In the interval $[-\pi, \pi)$, this triangular wave can be defined as Fig. 3.6(a)

$$H_{tri}^{ev}(t) = 2(1 - 2\frac{|t|}{\pi}), \quad |t| \leq \pi \tag{3.78}$$

the corresponding Fourier expansion is

$$H_{tri}^{ev}(t) = \frac{8}{\pi^2}[\cos t + \frac{\cos 3t}{3^2} + \frac{\cos 5t}{5^2} + \cdots] = \frac{8}{\pi^2} \sum_{n=1}^{\infty} \frac{\cos(2n-1)t}{(2n-1)^2} \tag{3.79}$$

Chebyshev had

$$\cos t = \frac{\pi^2}{8} \sum_{n=1}^{\infty} \mu(2n-1) \frac{H_{tri}^{ev}([2n-1]t)}{(2n-1)^2} \tag{3.80}$$

Note that the set of sine-cosine waves is orthogonal, but the set of triangular waves is not. Eq. (3.80) is a special case of Chebyshev's two theorems (see Chapter 1), that is,

Theorem 3.9.

$$g(t) = \sum_{n=1}^{\infty} \frac{f([2n-1]t)}{(2n-1)^\alpha} \quad \Leftrightarrow \quad f(t) = \sum_{n=1}^{\infty} \frac{\mu(2n-1)}{(2n-1)^\alpha} g([2n-1]t) \tag{3.81}$$

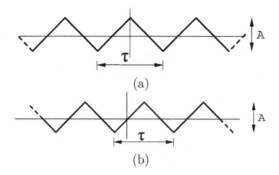

Fig. 3.6 Triangular waves: (a) Even and (b) odd

and

Theorem 3.10.

$$g(t) = \sum_{n=1}^{\infty} (-1)^{n+1} \frac{f([2n-1]t)}{(2n-1)^\alpha} \Leftrightarrow$$

$$f(t) = \sum_{n=1}^{\infty} (-1)^{n+1} \frac{\mu(2n-1)}{(2n-1)^\alpha} g([2n-1]t) \quad (3.82)$$

For the case of an odd triangular wave as Fig. 3.6(b)

$$H_{tri}^{od}(t) = \sum_{n=1}^{\infty} (-1)^{n+1} \frac{\sin(2n-1)t}{(2n-1)^2} \quad (3.83)$$

it corresponds to

$$\sin t = \sum_{n=1}^{\infty} (-1)^{n+1} \mu(2n-1) \frac{H_{tri}^{od}((2n-1)t)}{(2n-1)^2} \quad (3.84)$$

For the case of an even square wave as Fig. 3.7(a)

$$H_{sq}^{ev}(t) = \frac{4}{\pi} \sum_{n=1}^{\infty} (-1)^{n+1} \frac{\cos(2n-1)t}{(2n-1)} \quad (3.85)$$

it follows that

$$\cos t = \frac{\pi}{4} \sum_{n=1}^{\infty} (-1)^{n+1} \mu(2n-1) \frac{H_{sq}^{ev}((2n-1)t)}{(2n-1)} \quad (3.86)$$

For the case of an odd square wave as Fig. 3.7(b)

$$H_{sq}^{od}(t) = \frac{4}{\pi} \sum_{n=1}^{\infty} \frac{\sin(2n-1)t}{(2n-1)} \quad (3.87)$$

we have

$$\sin t = \frac{\pi}{4} \sum_{n=1}^{\infty} \mu(2n-1) \frac{H_{sq}^{od}((2n-1)t)}{(2n-1)} \qquad (3.88)$$

For a saw wave shown in Fig. 3.8, we have

$$H_{s1}^{od}(t) = \frac{2}{\pi} \sum_{n=1}^{\infty} (-1)^{n+1} \frac{\sin nt}{n} \qquad (3.89)$$

then

$$\sin t = \frac{\pi}{2} \sum_{n=1}^{\infty} (-1)^{n+1} \mu(n) \frac{H_{s1}^{od}(nt)}{n} \qquad (3.90)$$

Note that in the above both the Fourier expansion coefficient and the variable are multiplicative arithmetic functions, thus the inversion becomes simple. In addition, it is worth to note that the set of square waves, the set

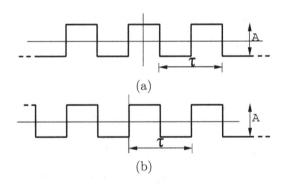

Fig. 3.7 Square waves with (a) even and (b) odd

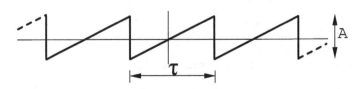

Fig. 3.8 Saw wave

of triangular waves, and so on are not an orthogonal set. For example,

$$\int_0^{2\pi} H_{sq}^{od}(x) H_{sq}^{od}(2x) dx$$

$$= \frac{4}{\pi} \frac{4}{\pi} \sum_{n=1}^{\infty} \sum_{m=1}^{\infty} \frac{1}{(2n-1)(2m-1)} \int_0^{2\pi} \sin(2n-1)x \sin(2m-1)dx$$

$$= \frac{16}{\pi^2} \sum_{n=1}^{\infty} \frac{1}{(2n-1)^2} = \frac{12}{\pi^2} \frac{\pi^2}{6} = 2 \neq 0.$$

3.5.2 From orthogonality to biorthogonality

First of all, let us introduce a modified Möbius series inversion formula

Theorem 3.11. If $h(1) \neq 0$, then

$$H(t) = \sum_{n=1}^{\infty} h(n) \sin nt \quad \Rightarrow \quad \sin t = \sum_{n=1}^{\infty} h^{-1}(n) H(nt) \quad (3.91)$$

where $\{h^{-1}(n)\}$ is the inverse of $\{h(n)\}$ satisfying

$$\sum_{n|k} h^{-1}(n) h(\frac{k}{n}) = \delta_{k,1} \quad (3.92)$$

the proof is omitted. Now, we consider an arbitrary odd periodic function $f(t)$ as

$$f(t) = \sum_{n=1}^{\infty} b(n) \sin nt. \quad (3.93)$$

The question is how to expand $f(t)$ by the functional set $\{H_n(t) \equiv H(nt)\}$. Based on Theorem 3.1 we have

$$f(t) = \sum_{n=1}^{\infty} b(n) \sum_{m=1}^{\infty} h^{-1}(m) H(mnt)$$

$$= \sum_{s=1}^{\infty} \left\{ \sum_{n|s} b(n) h^{-1}(\frac{s}{n}) \right\} H(st)$$

$$= \sum_{s=1}^{\infty} c(s) H_s(t). \quad (3.94)$$

Here, $c(s)$ is equal to the Dirichlet product of $\{b(n)\}$ and $\{h^{-1}(n)\}$, it is also the expansion coefficient of $f(t)$ in the non-orthogonal set $\{H_s(t)\}$. In other words,

$$c(s) = \sum_{n|s} b(n) h^{-1}(\frac{s}{n}) = \sum_{n|s} b(\frac{s}{n}) h^{-1}(n) = [b \otimes h^{-1}](s). \tag{3.95}$$

Note that it can be shown that $\lim\limits_{s \to \infty} c(s) \to 0$. In fact, from Eq. (3.91) it follows that

$$\sum_{n=1}^{\infty} |h(n)|^2 < \infty$$

due to finiteness of energy in a finite system. In similar, from Eq. (3.93) we have

$$\sum_{n=1}^{\infty} |b(n)|^2 < \infty \text{ or } \sum_{n=k}^{\infty} |b(n)|^2 \to 0 \text{ as } k \to \infty$$

Thus it follows that

$$\left| \int_{-\pi}^{\pi} f(t) H(kt) dt \right| = \pi \left| \sum_{n=1}^{\infty} b(nk) h(n) \right| \leq \pi \sqrt{\sum_{n=1}^{\infty} b^2(nk)} \sqrt{\sum_{n=1}^{\infty} h^2(n)}$$

$$\leq \pi \sqrt{\sum_{n=k}^{\infty} b^2(n)} \sqrt{\sum_{n=1}^{\infty} h^2(n)} \to 0 \text{ as } k \to \infty$$

Note that by using the law of conservation of energy, or a generalized Parseval's theorem, it can be directly given

$$\sum_{n=1}^{\infty} |c(n)|^2 < \infty$$

Based on the non-orthogonal expansion, one may explore the dual orthogonality in it. Considering that

$$\frac{1}{\pi} \int_0^{2\pi} \sin mt \, \sin nt \, dt = \delta_{m,n} \tag{3.96}$$

then we have

$$c(s) = \left\{ \sum_{n|s} b(n) h^{-1}(\frac{s}{n}) \right\}$$

$$= \sum_{n=1}^{\infty} \frac{1}{\pi} \int_0^{2\pi} \sin mt \, \sin nt \, dt \left\{ \sum_{m|s} b(n) h^{-1}(\frac{s}{m}) \right\}$$

$$= \frac{1}{\pi} \int_0^{2\pi} \left\{ \sum_{n=1}^{\infty} b(n) \sin nt \right\} \left\{ \sum_{m|s} h^{-1}(\frac{s}{m}) \sin mt \right\} dx$$

Therefore,

$$c(s) = \frac{1}{\pi} \int_0^{2\pi} dt\, f(t) \left\{ \sum_{m|s} h^{-1}(\frac{s}{m}) \sin mt \right\} \tag{3.97}$$

Now we can define a functional set $\{\widetilde{H}_s(t)\}$ as the Dirichlet product between the inverse of h and sine function, i.e.,

Definition 3.5.

$$\widetilde{H}_s(t) = \sum_{m|s} h^{-1}(\frac{s}{m}) \sin mt \tag{3.98}$$

thus the non-orthogonal expansion coefficient of $f(t)$ becomes

$$c(s) = \frac{1}{\pi} \int_0^{2\pi} dt\, f(t) \widetilde{H}_s(t) \tag{3.99}$$

Note that $\widetilde{H}_n(t) \neq \tilde{H}(nt)$ although $H_n(t) \equiv H(nt)$. It is easy to prove a very important dual-orthogonality between $\{H_n(t)\}$ and $\{\widetilde{H}_m(t)\}$. In fact,

$$\frac{1}{\pi} \int_0^{2\pi} H_m(t) \widetilde{H}_n(t)\, dt = \frac{1}{\pi} \int_0^{2\pi} \left\{ \sum_{k=1}^\infty h(k) \sin kmt \right\} \left\{ \sum_{j|n} h^{-1}(\frac{n}{j}) \sin jt \right\} dt$$

$$= \sum_{k=1}^\infty h(k) \cdot \delta_{km,j} \cdot \sum_{j|n} h^{-1}(\frac{n}{j}) = \sum_{j|n} h(\frac{j}{m}) h^{-1}(\frac{n}{j})$$

$$= \sum_{j|n} h(\frac{j}{m}) h^{-1}(\frac{n/m}{j/m}) = \delta_{\frac{n}{m},1} = \delta_{n,m}.$$

The dually orthogonal relation is given by

Theorem 3.12.

$$\frac{1}{\pi} \int_0^{2\pi} H_m(t) \widetilde{H}_n(t)\, dt = \delta_{n,m} \tag{3.100}$$

is just like the mutually reciprocal relation between the primitive vectors in the direct lattice and the basis vectors in the reciprocal lattice.

3.5.3 Multiplicative dual orthogonality and square wave representation

An even square wave with zero direct component can be expressed as

$$H(t) = H(t + 2n\pi) = \frac{\pi}{4} sgn(\cos t), \quad |t| \leq \pi \qquad (3.101)$$

where $sgn(t)$ is the unit sign function. The fourier expansion of $H(t)$ is

$$H_1(t) \equiv H(t) = \sum_{n=1}^{\infty} h(n) \cos nt \qquad (3.102)$$

where $h(2k-1) = (-1)^{k+1}/(2k-1)$ and $h(2k) = 0$. Obviously it is an unconditional convergent series[You80].
From the previous theorem, we have

$$h_\otimes^{-1}(2k-1) = (-1)^{k+1} \frac{\mu(2k-1)}{2k-1} \quad \text{and} \quad h_\otimes^{-1}(2k) = 0 \qquad (3.103)$$

This is equivalent to the recursive relation

$$\sum_{n|s} h_\otimes^{-1}(n) h(\frac{s}{n}) = \delta_{s,1}. \qquad (3.104)$$

As before, the dual functional sets can be written as

$$H_n(t) = \sum_{m=1}^{\infty} h(m) \cos(nm)t \qquad (3.105)$$

and

$$\widetilde{H}_n(t) = \sum_{m|n} h_\otimes^{-1}(\frac{n}{m}) \cos mt. \qquad (3.106)$$

Therefore,

$$\begin{aligned}
&\widetilde{H}_1(t) = \cos t, \quad \widetilde{H}_2(t) = \cos 2t, \\
&\widetilde{H}_3(t) = \cos 3t + \frac{1}{3}\cos t, \quad \widetilde{H}_4(t) = \cos 4t, \\
&\widetilde{H}_5(t) = \cos 5t - \frac{1}{5}\cos t, \quad \widetilde{H}_6(t) = \cos 6t + \frac{1}{3}\cos 2t + \cdots.
\end{aligned} \qquad (3.107)$$

Figure 3.9 shows a set of square waves $\{H_n(t)\}$ and the corresponding reciprocal functional set $\widetilde{H}_n(t)$. The biorthogonal relation

$$\frac{1}{\pi} \int_0^{2\pi} H_m(t) \widetilde{H}_n(t) \, dt = \delta_{m,n} \qquad (3.108)$$

can be checked directly. In other words, the set $\{H_n(t)\}$ and the set $\{\widetilde{H}_n(t)\}$ form a mutually reciprocal pair, or a dual pair. From the relation between

$\{H_n(t)\}$ and $\{\widetilde{H}_n(t)\}$ it can be concluded that any periodic function $f(t)$ can be expressed as a sum of square waves.

$$f(t) = \sum_{s=1}^{\infty} c(s) H(st) \qquad (3.109)$$

The coefficient can be evaluated based on

$$c(s) = \frac{1}{\pi} \int_0^{2\pi} dt\, f(t) \widetilde{H}_s(t) \qquad (3.110)$$

as before. This expression is much simpler than Walsh function in Fig. 3.10.

3.5.4 Multiplicative biorthogonal representation for saw waves

Now we are going to discuss the saw wave as

$$H(t) = H(t + 2n\pi) = \begin{cases} \frac{\pi - x}{\pi}, & 0 \leq t < \pi, \\ \frac{-\pi - x}{\pi}, & -\pi \leq t < 0. \end{cases} \qquad (3.111)$$

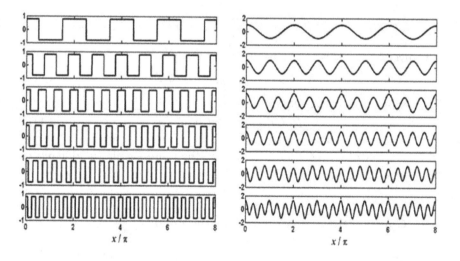

Fig. 3.9 A set of even square waves and its reciprocal set

Fig. 3.10 American Mathematician Walsh (1895–1973) and Walsh function

Corresponding Fourier expansion is

$$H_1(t) = \sum_{m=1}^{\infty} \frac{1}{m} \sin mt \tag{3.112}$$

and

$$H_n(t) = \sum_{m=1}^{\infty} \frac{1}{m} \sin nmt. \tag{3.113}$$

The inversion coefficient can be written directly as

$$h^{-1}(m) = \mu(m)a(m) = \frac{\mu(m)}{m} \tag{3.114}$$

The reciprocal function set is

$$\widetilde{H}_n(t) = \sum_{d|n} a_\otimes^{-1}(d) \sin \frac{n}{d} t \tag{3.115}$$

which is shown as Fig. 3.11. Obviously, there are significant differences between $\{\widetilde{H}_n(t)\}$ and $\{H_n(x)\}$.

3.6 Construction of Additive Biorthogonality

In the preceding section we have introduced the construction of multiplicative biorthogonality, and in this section we present the construction of additive biorthogonality.

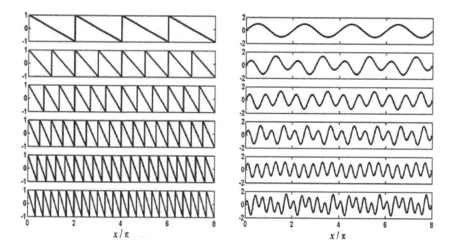

Fig. 3.11 Set of saw waves and its reciprocal set

3.6.1 Basic theorem on additively orthogonal expansion

Theorem 3.13. Let

$$\sum_{m+n=k} r^{-1}(n)r(m) = \delta_{k,0} \qquad (3.116)$$

then

$$H_m(t) = \sum_{n=0}^{\infty} r(n)\cos(m+n)t \quad \Leftrightarrow \quad \cos mt = \sum_{n=0}^{\infty} r^{-1}(n) H_{m+n}(t) \quad (3.117)$$

The proof is left for the reader. Now let us show how the orthogonal basis can change into a biorthogonal basis for a periodic function $f(t)$.

$$f(t) = \sum_{n=0}^{\infty} b(n)\cos nt = \sum_{n=0}^{\infty} b(n) \sum_{m=0}^{\infty} r^{-1}(m) H_{m+n}(t)$$

$$= \sum_{s=0}^{\infty} \{ \sum_{0 \le n \le s} r^{-1}(s-n) b(n) \} H_s(t) = \sum_{s=0}^{\infty} c(s) H_s(t). \qquad (3.118)$$

This is the non-orthogonal expansion of $f(t)$ on the set of $\{H_n(t)\}$, the expansion coefficient $c(s)$ is equal to the Dirichlet sum between $b(n)$ and $r(n)$, i.e.,

$$c(s) = \{b \oplus r\}(s) = \sum_{0 \leq n \leq s} r^{-1}(s-n)b(n) = \sum_{0 \leq n \leq s} r^{-1}(n)b(s-n) \quad (3.119)$$

where $b(n)$ is the orthogonal expansion coefficient of $f(t)$, $r(n)$ is the orthogonal expansion coefficient of $H_m(t)$. Note that

$$\frac{1}{\pi}\int_0^{2\pi} \cos mt \cos nt\, dt = \delta_{m,n}[1 + \delta_{n,0}] \quad (3.120)$$

hence,

$$c(s) = \sum_{0 \leq n \leq s} r^{-1}(s-n)b(n) = \sum_{0 \leq n \leq s} r^{-1}(s-n) \sum_{m=0}^{\infty} \delta_{m,n} b(m)$$

$$= \sum_{0 \leq n \leq s} r^{-1}(s-n) \sum_{m=0}^{\infty} \frac{1}{(1+\delta_{m,0})\pi} \int_0^{2\pi} \cos mt \cos nt\, b(m)\, dt$$

$$= \frac{1}{(1+\delta_{m,0})\pi} \int_0^{2\pi} dt \{\sum_{m=0}^{\infty} b(m) \cos mt\}\{\sum_{0 \leq n \leq s} r^{-1}(s-n) \cos nt\}$$

$$= \frac{1}{(1+\delta_{m,0})\pi} \int_0^{2\pi} dt\, f(t)\, \widetilde{H}_s(t) \quad (3.121)$$

In fact, this shows the biorthogonal relationship between $\widetilde{H}_s(t)$ and $H_\ell(t)$:

Theorem 3.14.

$$\frac{1}{(1+\delta_{m,0})\pi} \int_0^{2\pi} dt\, H_\ell(t)\, \widetilde{H}_s(t) = \delta_{\ell,s}. \quad (3.122)$$

Proof.

$$\frac{1}{(1+\delta_{\ell,0})\pi} \int_0^{2\pi} dt\, H_\ell(t)\, \widetilde{H}_s(t)$$

$$= \frac{1}{(1+\delta_{\ell,0})\pi} \int_0^{2\pi} dt\{\sum_{n=0}^{\infty} r(n) \cos(\ell+n)t\}\{\sum_{0 \leq j \leq s} r^{-1}(s-j) \cos jt\}$$

$$= \sum_{n=0}^{\infty} r(n)\delta_{\ell+n,j} \sum_{0 \leq j \leq s} r^{-1}(s-j) = \sum_{0 \leq j \leq s} r^{-1}(s-j)r(j-\ell)$$

$$= \sum_{0 \leq j-\ell \leq s-\ell} r^{-1}[(s-\ell)-(j-\ell)]r(j-\ell) = \delta_{s-\ell,0} = \delta_{s,\ell}. \qquad \square$$

3.6.2 Derivative biorthogonality from even square waves

From the square wave

$$1 + \cos t - \frac{1}{3}\cos 3t + \frac{1}{5}\cos 5t - \frac{1}{7}\cos 7t + \cdots \qquad (3.123)$$

we construct or derive a set of functions as

$$H_0(t) = \frac{1}{2} + \cos t - \frac{1}{3}\cos 3t + \frac{1}{5}\cos 5t - \cdots$$

$$H_1(t) = \cos t + \cos 2t - \frac{1}{3}\cos 4t + \frac{1}{5}\cos 6t - \cdots$$

$$H_2(t) = \cos 2t + \cos 3t - \frac{1}{3}\cos 5t + \frac{1}{5}\cos 7t - \cdots$$

$$H_3(t) = \cos 3t + \cos 4t - \frac{1}{3}\cos 6t + \frac{1}{5}\cos 8t - \cdots$$

$$H_4(t) = \cos 4t + \cos 5t - \frac{1}{3}\cos 7t + \frac{1}{5}\cos 9t - \cdots$$

$$H_5(t) = \cos 5t + \cos 6t - \frac{1}{3}\cos 8t + \frac{1}{5}\cos 10t - \cdots$$

$$\cdots\cdots$$

$$(3.124)$$

The corresponding reciprocal set $\{G_i(t)\}$ or $\{\widetilde{H_i}(t)\}$ is

$$G_0(t) = 1$$

$$G_1(t) = \cos t - 1$$

$$G_2(t) = \cos 2t - \cos t + 1$$

$$G_3(t) = \cos 3t - \cos 2t + \cos t - \frac{2}{3}$$

$$G_4(t) = \cos 4t - \cos 3t + \cos 2t - \frac{2}{3}\cos t + \frac{1}{3}$$

$$G_5(t) = \cos 5t - \cos 4t + \cos 3t - \frac{2}{3}\cos 2t + \frac{1}{3}\cos t - \frac{1}{5}$$

$$\cdots$$

$$(3.125)$$

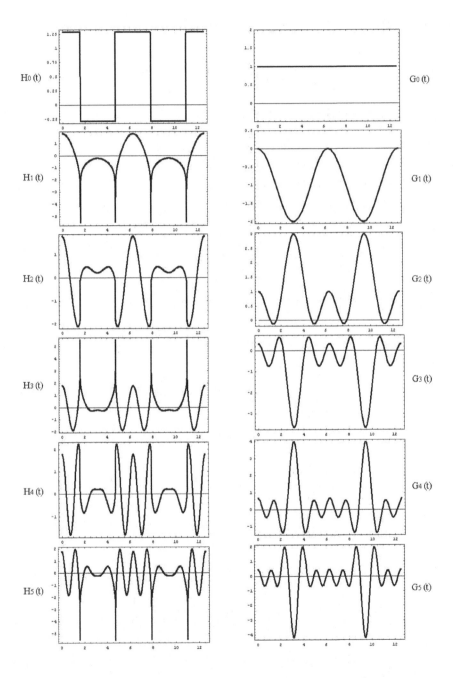

Fig. 3.12 Derivative set of square waves and the reciprocal set

The patterns in Fig. 3.12 look much more complicated than those in the multiplicative cases. Sometimes this complexity is very useful when we can control it.

Now let us consider a little more complicated case. Replacing the $\cos nt$ in the above equations, by $P_n(\cos t)$. In other words, Eq. (3.123) is replaced by

$$1 + P_1(\cos t) - \frac{1}{3}P_3(\cos t) + \frac{1}{5}P_5(\cos t) - \frac{1}{7}P_7(\cos t) + \cdots \quad (3.126)$$

then the Legendre derivative set becomes

$$H_0(t) = \frac{1}{2} + P_1(\cos t) - \frac{1}{3}P_3(\cos t) + \frac{1}{5}P_5(\cos t) - \cdots$$

$$H_1(t) = P_1(\cos t) + P_2(\cos t) - \frac{1}{3}P_4(\cos t) + \frac{1}{5}P_6(\cos t) - \cdots$$

$$H_2(t) = P_2(\cos t) + P_3(\cos t) - \frac{1}{3}P_5(\cos t) + \frac{1}{5}P_7(\cos t) - \cdots$$

$$H_3(t) = P_3(\cos t) + P_4(\cos t) - \frac{1}{3}P_6(\cos t) + \frac{1}{5}P_8(\cos t) - \cdots \quad (3.127)$$

$$H_4(t) = P_4(\cos t) + P_5(\cos t) - \frac{1}{3}P_7(\cos t) + \frac{1}{5}P_9(\cos t) - \cdots$$

$$H_5(t) = P_5(\cos t) + P_6(\cos t) - \frac{1}{3}P_8(\cos t) + \frac{1}{5}P_{10}(\cos t) - \cdots$$

$$\ldots\ldots$$

$$G_0(t) = 1$$
$$G_1(t) = P_1(\cos t) - 1$$
$$G_2(t) = P_2(\cos t) - P_1(\cos t) + 1 \quad (3.128)$$
$$G_3(t) = P_3(\cos t) - P_2(\cos t) + P_1(\cos t) - \frac{2}{3}$$
$$G_4(t) = P_4(\cos t) - P_3(\cos t) + P_2(\cos t) - \frac{2}{3}P_1(\cos t) + \frac{1}{3}$$
$$G_5(t) = P_5(\cos t) - P_4(\cos t) + P_3(\cos t) - \frac{2}{3}P_2(\cos t) + \frac{1}{3}P_1(\cos t) - \frac{1}{5}$$
$$\ldots$$

The Legendre derivative functional set[4] and the corresponding reciprocal functional set are shown in Fig. 3.14, in which the biorthogonal relation

---
[4]Legendre was given a top quality education in mathematics and physics. From 1775 to 1780 his actual task was to determine the curve described by cannonballs and bombs, taking into consideration the resistance of the air; give rules for obtaining the ranges corresponding to different initial velocities and to different angles of projection. Legendre next studied the attraction of ellipsoids. He then introduced what we call today the

becomes

$$\int_0^{2\pi} H_m(\cos mt)G_n(\cos nt)\sin t\, dt = \delta_{m,n} \qquad (3.129)$$

Similarly, we can construct the derivative set of Bessel functions[5].

Fig. 3.13   A.M. Legendre (1752–1833) and F.W. Bessel (1784–1846)

Legendre functions to determine, using power series, the attraction of an ellipsoid at any exterior point. His results were highly praised by Laplace, thus he was appointed an adjoint in the Academie des Sciences. In 1824 Legendre refused to vote for the government's candidate for the Institut Nationalas a result, his pension was stopped and he died in poverty.

[5] In January 1799, at the age of 14, Bessel left school to become an apprentice to the commercial firm of Kulenkamp in Bremen. The firm was involved in the import-export business. His interests turned towards the problem of finding the position of a ship at sea. This in turn led him to study astronomy and mathematics and he began to make observations to determine longitude. In 1809, at the age of 26, Bessel was appointed director of Frederick William III of Prussia's new Königsberg Observatory and professor of astronomy. Bessel functions appear as coefficients in the series expansion of the indirect perturbation of a planet, that is the motion caused by the motion of the Sun caused by the perturbing body. Bessel also had a very significant impact on university teaching despite the fact that he never had a university education. Klein describes how the name of Bessel, together with the names of Jacobi and Franz Neumann, is intimately linked to the reform of teaching at universities, first in Germany and then throughout the world.

Inverse Problems in Fermion Systems 111

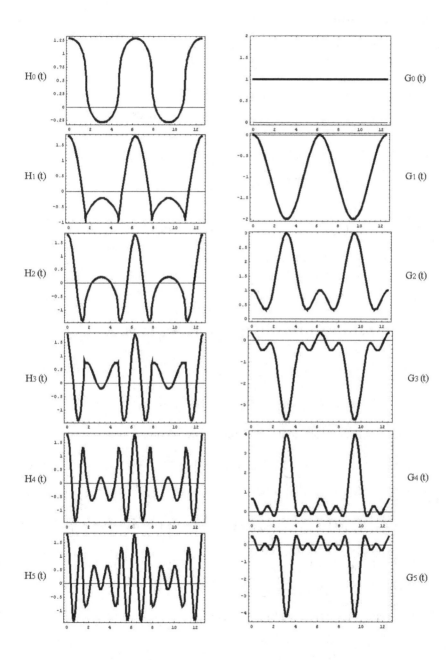

Fig. 3.14 Derivative set of Legendre functions and the reciprocal set

3.6.3 Derivative set from triangular wave

Let us consider the functional set $\{H_i(t)\}$ as

$$H_0(t) = \sin t + \sum_{k=1}^{\infty} \frac{1}{k} \cos kt$$

$$= \sin t + \cos t + \frac{1}{2}\cos 2t + \frac{1}{3}\cos 3t + \cdots \quad (3.130)$$

and

$$H_j(t) = \cos jt + \sum_{k=j}^{\infty} \frac{1}{k+1-j}\cos(k+1)t \quad \text{where} \quad j \geq 1 \quad (3.131)$$

That is

$$H_0(t) = \sin t + \cos t + \frac{1}{2}\cos 2t + \frac{1}{3}\cos 3t + \cdots$$

$$H_1(t) = \cos t + cos2t + \frac{1}{2}\cos 3t + \frac{1}{3}\cos 4t + \cdots$$

$$H_2(t) = \cos 2t + \cos 3t + \frac{1}{2}\cos 4t + \frac{1}{3}\cos 5t + \cdots$$

$$H_3(t) = \cos 3t + \cos 4t + \frac{1}{2}\cos 5t + \frac{1}{3}\cos 6t + \cdots$$

$$\cdots$$

$$H_6(t) = \cos 6t + \cos 7t + \frac{1}{2}\cos 8t + \frac{1}{3}\cos 9t + \cdots$$

$$\cdots \quad (3.132)$$

The inversion coefficient can be obtained by recursive relation

$$\sum_{n=0}^{k} a(k-n)a^{-1}(n) = \delta_{k,0} \quad (3.133)$$

therefore,

$$a^{-1}(1) = 1,\ a^{-1}(2) = -1,\ a^{-1}(3) = \frac{1}{2},\ a^{-1}(4) = \frac{-1}{3},\ a^{-1}(5) = \frac{1}{6},$$

$$a^{-1}(6) = \frac{-7}{60},\ a^{-1}(7) = \frac{19}{360},\ a^{-1}(8) = \frac{-3}{70},$$

$$a^{-1}(9) = \frac{5}{336},\ a^{-1}(10) = \frac{-13}{756},\ \cdots$$

The corresponding reciprocal functions are

$$G_\ell(t) = \begin{cases} a^{-1}(0)\sin t & \text{if} \quad \ell = 0 \\ \sum_{k=0}^{\ell} a^{-1}(\ell - k)\cos kt & \text{if} \quad \ell \geq 1 \end{cases} \quad (3.134)$$

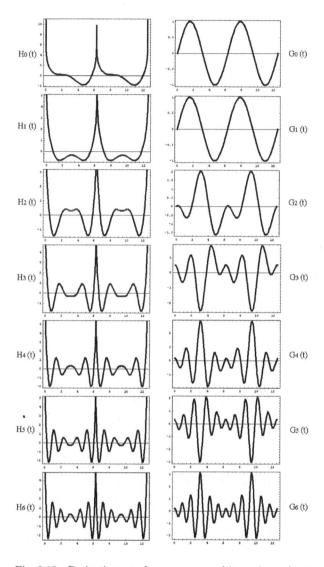

Fig. 3.15 Derivative set of a saw wave and its reciprocal set

In other words,
$$G_0(t) = \sin t$$
$$G_1(t) = \cos t - \sin t$$
$$G_2(t) = \cos 2t - \cos t + \frac{1}{2}\sin t$$

$$G_3(t) = \cos 3t - \cos 2t + \frac{1}{2}\cos t - \frac{1}{3}\sin t$$
$$G_4(t) = \cos 4t - \cos 3t + \frac{1}{2}\cos 2t - \frac{1}{3}\cos t + \frac{1}{6}\sin t$$
$$G_5(t) = \cos 5t - \cos 4t + \frac{1}{2}\cos 3t - \frac{1}{3}\cos 2t + \frac{1}{6}\cos t - \frac{7}{60}\sin t$$
$$G_6(t) = \cos 6t - \cos 5t + \frac{1}{2}\cos 4t - \frac{1}{3}\cos 3t$$
$$+ \frac{1}{6}\cos 2t - \frac{7}{60}\cos t + \frac{19}{360}\sin t$$
$$\cdots \quad (3.135)$$

The biorthogonality between $\{H_i(t)\}$ and $\{G_j(t)\}$ can be checked directly.

3.6.4 Another derivative set by saw wave

Now let us show the another derivative set from a saw wave.
$$H_i(t) = \sum_{k=i}^{\infty} \frac{1}{k-i+1}\sin kt \quad i = 1, 2, 3, \ldots \quad (3.136)$$
In other words,
$$H_1(t) = \sin t + \frac{1}{2}\sin 2t + \frac{1}{3}\sin 3t + \cdots$$
$$H_2(t) = \sin 2t + \frac{1}{2}\sin 3t + \frac{1}{3}\sin 4t + \cdots$$
$$H_3(t) = \sin 3t + \frac{1}{2}\sin 4t + \frac{1}{3}\sin 5t + \cdots \quad (3.137)$$
$$\cdots$$
$$H_6(t) = \sin 6t + \frac{1}{2}\sin 7t + \frac{1}{3}\sin 8t + \cdots$$

As before, the coefficient $a^{-1}(n)$ satisfies
$$\sum_{m=1}^{k} a^{-1}(m)a(k-m+1) = \delta_{k1}$$
or
$$a^{-1}(1)a(1) = 1$$
$$a^{-1}(1)a(2) + a^{-1}(2)a(1) = 0$$
$$a^{-1}(1)a(3) + a^{-1}(2)a(2) + a^{-1}(3)a(1) = 0$$
$$a^{-1}(1)a(4) + a^{-1}(2)a(3) + a^{-1}(3)a(2) + a^{-1}(4)a(1) = 0$$
$$\cdots$$

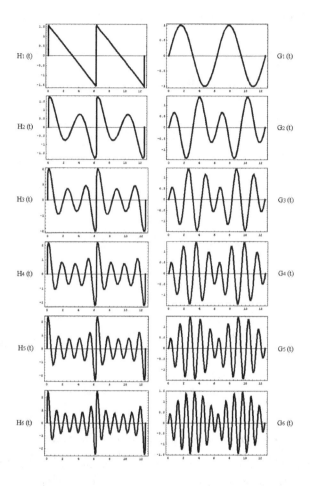

Fig. 3.16 Another derivative set from saw wave and its reciprocal set

the concrete coefficients are given that

$$a^{-1}(1) = 1 \qquad a^{-1}(2) = \frac{-1}{2}$$

$$a^{-1}(3) = \frac{-1}{12} \qquad a^{-1}(4) = \frac{-1}{24}$$

$$a^{-1}(5) = -19/720 \qquad a^{-1}(6) = \frac{-3}{160}$$

$$a^{-1}(7) = \frac{-863}{60480} \qquad a^{-1}(8) = \frac{-275}{24192}$$

$$a^{-1}(9) = \frac{-33953}{3628800} \qquad a^{-1}(10) = \frac{-8183}{1036800}$$

the reciprocal functions are shown in Fig. 3.16.

$$G_1(t) = \sin t$$
$$G_2(t) = -\frac{1}{2}\sin t + \sin 2t$$
$$G_3(t) = -\frac{1}{2}\sin t - \frac{1}{2}\sin 2t + \sin 3t$$
$$G_4(t) = -\frac{1}{24}\sin t - \frac{1}{12}\sin 2t - \frac{1}{2}\sin 3t + \sin 4t \quad (3.138)$$
$$G_5(t) = -\frac{19}{720}\sin t - \frac{1}{24}\sin 2t$$
$$\qquad -\frac{1}{12}\sin 3t - \frac{1}{2}\sin 4t + \sin 5t$$
$$G_6(t) = -\frac{3}{160}\sin t - \frac{19}{720}\sin 2t - \frac{1}{24}\sin 3t$$
$$\qquad -\frac{1}{12}\sin 4t - \frac{1}{2}\sin 5t + \sin 6t$$
$$\cdots$$

Combining these, we have

$$G_\ell(t) = \sum_{k=1}^{\ell} a^{-1}(\ell - k) \sin kt. \quad (3.139)$$

3.6.5 Biorthogonal modulation in communication

Usually an orthogonal set is used as the modulation function $\{M_i(t)\}$, while the same functional set is taken as demodulation $\{D_j(t)\}$.

$$D_i(t) = M_i(t) \quad (3.140)$$

$$\int_0^{2\pi} M_i(t) D_j(t) dt = \delta_{i,j} \quad (3.141)$$

For example,

$$D_i(t) = M_i(t) = \sin \omega t, \ \sin \omega 2t, \ \sin \omega 3t, \ \sin \omega 4t, \ \cdots. \quad (3.142)$$

Now we consider the biorthogonality, in which Eq. (3.140) are not available,

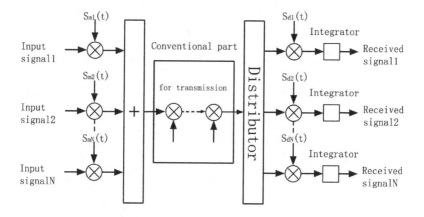

Fig. 3.17 Principle of multi-channel communication

and Eq. (3.142) are not available. In other words,

$$D_i(t) \neq M_i(t)$$

$$\int_0^{2\pi} M_i(t)M_j(t)dt \neq \delta_{i,j}$$

$$\int_0^{2\pi} D_i(t)D_j(t)dt \neq \delta_{i,j} \tag{3.143}$$

However, the biorthogonality leads

$$\int_0^{2\pi} M_i(t)D_j(t)dt = \delta_{i,j} \tag{3.144}$$

In Fig. 3.18, (a) shows the seven input signals and the corresponding modulation functions, (b) shows the products of the modulation function and the input signal one by one. The rule of the multiplier is to combine the input of signal and modulation into a output of the modulated signal. In part (a) of Fig. 3.19, the seven modulated signals are mixed together as the output of the modulator. Part (b) of Fig. 3.19 shows the first patterns of the seven demodulated signals. After filtering they become the

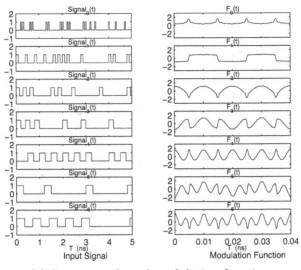

(a) Seven signals and modulation functions

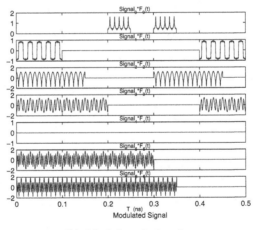

(b) Modulated signals

Fig. 3.18 The first step of modulation

recovered patterns as shown in the right part of Fig. 3.20. The matching demodulation functions are shown in the left part of Fig. 3.20. Finally, by pattern recognition and reshaping, the output signals should be quite closed to the input.

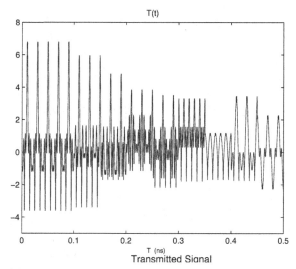

(a) Superposition: the second step of modulation

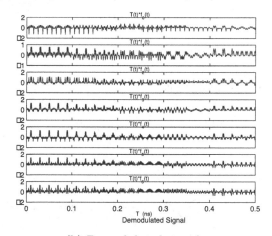

(b) Demodulated signals

Fig. 3.19 The first step for demodulation

3.7 Cesàro Inversion Formula of the Second Kind

In the last subsection, we have introduced the multiplicative Möbius-Cesàro inversion formula. In this section we discuss the Möbius-Cesàro inversion formula of the second kind. Ignoring the convergence problem as before,

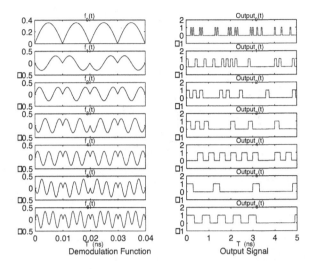

Fig. 3.20 Filtering: the second step of demodulation

the theorem can be expressed as

Theorem 3.15. Assuming that

$$g_m(g_n(x)) = g_{m+n}(x) \tag{3.145}$$

is hoed for any integers m and n, then

$$F(x) = \sum_{n=0}^{\infty} r(n)f(g_n(x)) \quad r(0) \neq 0 \Rightarrow$$
$$f(x) = \sum_{n=0}^{\infty} r_{\oplus}^{-1}(n)F(g_n(x)) \tag{3.146}$$

where

$$\sum_{n=0}^{k} r_{\oplus}^{-1}(n)r(k-n) = \delta_{k,0} \tag{3.147}$$

Proof.

$$\sum_{n=0}^{\infty} a_{\oplus}^{-1}(n) F(g_n(x)) = \sum_{n=1}^{\infty} a_{\oplus}^{-1}(n) \sum_{m=1}^{\infty} a(m) f(g_m[g_n(x)])$$

$$= \sum_{n=1}^{\infty} [\sum_{n=0}^{k} a_{\oplus}^{-1}(n) a(k-n)] f(g_k(x))$$

$$= \sum_{n=1}^{\infty} \delta_{k,0} f(g_k(x)) = f(g_0(x)) = f(x).$$

$\square$

For constructing $g_n(x)$, it is useful to know that the condition Eq. (3.145) can be expressed as

$$g_n(x) = F^{-1}(n + F(x)). \qquad (3.148)$$

Note that $F(x)$ is an arbitrary function, and $F^{-1}F(x) = x$, then

$$g_0(x) = F^{-1}(n + F(x))|_{n=0} = F^{-1}(F(x)) = x$$

and

$$g_m(g_n(x)) = g_m(F^{-1}[n + F(x)])$$
$$= F^{-1}(m + F\{F^{-1}(n + F(x))\})$$
$$= F^{-1}(m + [n + F(x)])$$
$$= F^{-1}[m + n + F(x)] = g_{m+n}(x).$$

There are many examples of $g_n(x)$ which satisfy $g_m[g_n(x)] = g_{m+n}(x)$. For example,

$$\begin{cases} g_n(x) = \sqrt{n + x^2} \\ \\ g_n(x) = (n + \sqrt{x})^2 \\ \\ g_n(x) = \ln(n + e^x) \\ \\ g_n(x) = \exp(n + \ln x) \\ \\ g_n(x) = \sin(n + \sin^{-1} x) \\ \\ g_n(x) = \sin^{-1}(n + \sin x) \end{cases} \qquad (3.149)$$

Obviously, if we changed n into $b(n)$ with the condition of $b(m) + b(n) = b(m+n)$, then much more modifications of the Cesàro theorem of second kind can be given.

The so called additive Cesàro inversion formulas can be outlined as follows.

$$g_m[g_n(x)] = g_{mn}(x) \;\;\Rightarrow\;\; g_n(x) = F^{-1}[F(x) + n],$$

$$F[g_n(x)] = \sum_{d=0}^{n} r(d) f[g_{n-d}(x)] \;\;\Leftrightarrow\;\; f[g_n(x)] = \sum_{d=0}^{n} r^{-1}(d) F[g_{n-d}(x)],$$

$$F(x) = F[g_0(x)] = \sum_{d=\infty}^{n} r(d) f[g_d(x)] \;\;\Leftrightarrow\;\; f(x) = \sum_{d=0}^{n} r^{-1}(d) F[g_d(x)],$$

with relation

$$\sum_{d=0}^{n} r^{-1}(d) r(n-d) = \delta_{n,1}.$$

3.8 Summary

Chapter 3 presented the Möbius inversion formula on the additive semi-group and its applications to fermion-like system. Several distinct aspects were addressed.

(1) The applications of the additive Möbius inversion formula to the Fourier deconvolution. Note that the series form solution is easy for approximation analysis and that either the fermion integral equation, emphasized here, or the boson integral equation can be considered as a convolution; to solve either one is equivalent to taking deconvolution.

(2) An additive Cesàro inversion formula or the Cesàro inversion formula of the second kind were introduced in detail[6].

(3) The applications of the additive Möbius inversion formula to inverse

---

[6] For practical convenience, some non-conventional terminologies were introduced in this chapter, such as the additive Möbius and Cesàro inversion formulas. From the viewpoint of the partially ordered set theory, which will be presented in Chapter 6, these terminologies feel natural.

fermion-like systems, including the integral equation of fermion system, the relaxation spectrum problem, and the surface absorption problem were presented. Note that the inverse boson system problems are related to the multiplicative semigroup, and the inverse fermion system problems are related to the additive semigroup.

(4) A $\delta - kernel$ method was presented.

(5) A systematic procedure for constructing biorthogonal functional set was applied to designing non-orthogonal modulation systems in signal processing.

(6) The mathematical details about the adsorption integral equation can be found in some literatures. In fact, the problem is known as ill-posed. The difficulty in solving the equation arises from the fact that small variations in the total isotherm Θ_t may lead to significant changes in the distribution function. Because Θ_t is usually given only as a discrete set of experimental data (with inevitable experimental errors) in limited range of pressures, there are much more other problems involved. According to [Bus2001], the method for solving this problem can be approximately classified into three kinds:

(a) Analytical methods first introduced by Sips[Sip48; Sip50] and soon after developed by Misra[Mis70] based on the integral transform theory. This approach assumes that the local isotherm an d the global one are given in an explicit analytical form.

(b) Numerical methods based on an elementary optimization of a parameterized form or more accurate methods, which take into account the ill-posed character of the Fredholm equation of first kind. The examples using regularization method can be found[Hou78; Vzo92].

(c) Approximate methods based on approximations imposed on the local isotherm. The most popular and commonly used approximation method is the condensation approximationHar68. There are also hybrid analytical-approximate methods like the method proposed by Villierras[Vil97]. The approach in this book assumes that the local isotherm and the global (or total) one are given in an explicit analytical form[Ram2000; Rus2001].

Chapter 4
Arithmetic Fourier Transform

4.1 Concept of Arithmetic Fourier Transform

From ancient optics, acoustics, and electrical engineering to the contemporary communication and information science, Fourier transform plays a non-fungible role[Bea75; Cha73], just as Joseph Fourier predicted, "Mathematics compares the most diverse phenomena and discovers the secret analogies that unite them". In Chapter 3, we presented the design of biorthogonal modulation by using generalized Möbius inversion formulas for periodic functions. This chapter will present a "design" of parallel algorithm of Fourier transform by using Möbius inverse formula too. Both Arithmetic

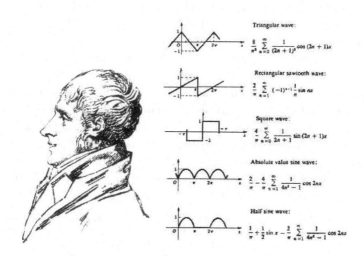

Fig. 4.1 Joseph Fourier (1768–1830) and Fourier series.

(a) (b)

Fig. 4.2 (a) James W. Cooley (1926–). (b) John W. Tukey (1915–2000)

Fourier transform (AFT) and Fast Fourier transform (FFT)[Coo65][1] are algorithms for discrete Fourier transform (DFT). The advantage of AFT regarding FFT is that except a few multiplications for calculating inverse of some integers, only the additive operations are involved in AFT. This is convenient for parallel algorithm, which has attracted great attention in signal processing, and very large scale integrated circuit (VLSI).

As early as 1903, German mathematician H. Bruns[Bru03] proposed a method for calculating Fourier coefficients of periodic functions by using Möbius inverse formula. Forty years later, in 1947, an American mathematician, Wintner, presented a similar algorithm[Win47], which is only available for even periodic functions. After another 40 years, in 1988, Tufts and Sadasiv[Tuf88] almost repeated Wintner's algorithm, but proposed its applications to signal processing and VLSI, and denoted it as AFT. Soon after, the Wintner's method was generalized by Schiff[Sch88] so that the coefficients for both sine and cosine can be calculated. In 1990, Reed introduced the AFT with a finite number of coefficients for a general periodic function[Ree90], and checked the algorithm of Bruns. Reed et al. discovered that Bruns' algorithm has not only generality but also simplification[Sch92; Hsu94]. Currently, either the Wintner's method or the Bruns' algorithm have been applied to Fourier convolution, Z-transform and design of VLSI. This chapter will review the principle of AFT based on

---

[1]Proposed by Cooley (1926–) and Tukey (1915–2000). Cooley is an American mathematician. He worked on quantum mechanical computations at the Courant Institute, New York University, from 1956 to 1962, when he joined the Research Staff at the IBM Watson Research Center, Yorktown Heights, NY. (He retired from IBM in 1991). Tukey was an American statistician. [From Wikipedia]

Möbius inverse formula. In order to overcome the sampling with nonequal time intervals, we propose a new way to get a clear understanding of the nonequal sampling in AFT based on the Ramannujan sum rule[Che97]. The main disadvantages of the current AFT is over-sampling, and the much larger work load for sine components.

4.2 Fundamental Theorem of AFT (Wintner)

4.2.1 Statement of the Wintner theorem

Theorem 4.1. Assume a periodic function $f(x)$ with vanished direct component

$$f(x) = \sum_{n=1}^{\infty} [a(n) \cos nx + b(n) \sin nx], \quad x \in (-\infty, \infty), \quad (4.1)$$

then

$$a(n) = \sum_{m=1}^{\infty} \frac{\mu(m)}{mn} \sum_{s=1}^{mn} f(2\pi \frac{s}{mn}) \quad (4.2)$$

and

$$b(n) = (-1)^k \sum_{m=1}^{\infty} \frac{\mu(m)}{mn} \sum_{s=1}^{mn} f(2\pi \frac{s}{mn} + 2\pi \frac{1}{2^{q+2}}), \quad (4.3)$$

where $n = 1, 2, \cdots$, and define k, q as

$$n = 2^q(2k+1), \quad k, q = 0, 1, 2, \ldots \quad (4.4)$$

Note that k, q are determined by n uniquely as in Table 4.1.

4.2.2 Proof of Eq. (4.2)

Proof. Set $x = 2\pi, \frac{2\pi}{2}, \frac{3\pi}{2}, \frac{4\pi}{2}, \ldots, \frac{2\pi}{n}$ in Eq. (4.1), it is given by

$$\sum_{r=1}^{\infty} a(r) = f(0) = f(2\pi) = S(1)$$

$$\sum_{r=1}^{\infty} a(2r) = \frac{1}{2}[f(\pi) + f(2\pi)] = S(2)$$

......

$$\sum_{r=1}^{\infty} a(nr) = \frac{1}{n} \sum_{s=1}^{n} f(2\pi \frac{s}{n}) = S(n). \quad (4.5)$$

Table 4.1 Uniqueness of k and q

$n = 2^q(2k+1)$	q	k	$2^{-(q+2)}$
1	0	0	1/4
2	1	0	1/8
3	0	1	1/4
4	2	0	1/16
5	0	2	1/4
6	1	1	1/8
7	0	3	1/4
8	3	0	1/32

where $S(n)$ represents the average of n sampling values with zero phase shifts. Equation (4.5) can be interpreted as follows.

$$\frac{1}{n}\sum_{s=1}^{n} f(2\pi\frac{s}{n}) = \frac{1}{n}\sum_{s=1}^{n}\sum_{k=1}^{\infty}[a_k \cos 2\pi\frac{ks}{n} + b_k \sin 2\pi\frac{ks}{n}]$$

$$= \sum_{k=1}^{\infty} a_k[\frac{1}{n}\sum_{s=1}^{n}\cos 2\pi\frac{ks}{n}] + \sum_{k=1}^{\infty} b_k[\frac{1}{n}\sum_{s=1}^{n}\sin 2\pi\frac{ks}{n}]$$

$$= \sum_{k} a_k|_{k=nr} = \sum_{r=1}^{\infty} a_{nr}$$

in which we have used the following two formulas:

$$\frac{1}{n}\sum_{m=1}^{n} \cos(\frac{2\pi km}{n}) = \begin{cases} 1, & \text{if } n|k \\ 0, & \text{if } n \nmid k \end{cases} \quad (4.6)$$

and

$$\frac{1}{n}\sum_{m=1}^{n} \sin(\frac{2\pi km}{n}) = 0, \text{ for all the integers k and n.} \quad (4.7)$$

The above two relations come from the equality

$$\frac{1}{n}\sum_{m=1}^{n} e^{i2\pi km/n} = \begin{cases} 1, & \text{if } n|k \quad \text{or} \quad k=rn \\ 0, & \text{if } n \nmid k \end{cases}. \quad (4.8)$$

Equation (4.8) can be obtained by geometric series directly. Based on Eq. (4.5) and the Möbius inverse formula, it is given by

$$a(n) = \sum_{m=1}^{\infty} \mu(m)S(mn) = \sum_{m=1}^{\infty} \frac{\mu(m)}{mn}\sum_{s=1}^{mn} f(2\pi\frac{s}{mn}). \quad (4.9)$$

This is essentially the same as Eq. (4.2). Note that in the above procedure the commutativity of summations has been used. □

4.2.3 Proof of Eq. (4.3)

Proof. For a certain integer $n = 2^q(2k+1)$, we introduce a new variable x such that

$$x = t + \frac{2\pi}{2^{q+2}} \equiv t + \alpha. \tag{4.10}$$

Substitute Eq. (4.10) for Eq. (4.1), then

$$f(t+\alpha) = \sum_{n=1}^{\infty}[a(n)\cos n(t+\alpha) + b(n)\sin n(t+\alpha)]$$

$$= \sum_{n=1}^{\infty}[a(n)(\cos nt \cos n\alpha - \sin nt \sin n\alpha)]$$

$$+ \sum_{n=1}^{\infty}[b(n)(\sin nt \cos n\alpha + \cos nt \sin n\alpha)].$$

Change the expression as

$$\sum_{n=1}^{\infty}[a'(n)\ \cos nt + b'(n)\ \sin nt] = f(t + \frac{2\pi}{2^{q+2}}) \tag{4.11}$$

in which

$$a'(n) = a(n)\cos n\alpha + b(n)\sin n\alpha \tag{4.12}$$

while

$$b'(n) = -a(n)\ \sin n\alpha + b(n)\ \cos n\alpha \tag{4.13}$$

By using $n = 2^q(2k+1)$ and comparing Eqs. (4.13) and (4.12), the relationships among coefficients $a(n), b(n), a'(n), b'(n)$ are

$$a'(n) = (-1)^k b(n), \qquad b'(n) = (-1)^{k+1} a(n) \tag{4.14}$$

Considering Eqs. (4.9), (4.11) and (4.14), we have

$$b(n) = (-1)^k a'(n) == (-1)^k \sum_{m=1}^{\infty} \frac{\mu(m)}{mn} \sum_{s=1}^{mn} f(2\pi \frac{s}{mn} + 2\pi \frac{1}{2^{q+2}}) \tag{4.15}$$

Thus, Eq. (4.3) is proven. □

By the way, in Eq. (4.15) one may denote $S(mn, \alpha)$ as

$$S(mn, \alpha) = \frac{1}{mn}\sum_{s=1}^{mn} f(2\pi \frac{s}{mn} + 2\pi\alpha) \tag{4.16}$$

where

$$\alpha = \frac{1}{2^{q+2}} \tag{4.17}$$

Then Eq. (4.15) becomes

$$(-1)^k b(n) = \sum_{m=1}^{\infty} \mu(m) S(mn, \alpha) \qquad (4.18)$$

From the Möbius inverse formula we have

$$S(n, \alpha) = \sum_{m=1}^{\infty} (-1)^k b(mn) \qquad (4.19)$$

After Wintner, 40 years later, Tufts and Sadasiv also obtained the above result independently[Tuf88]. Note that, different from ordinary discrete Fourier transform, in Eqs. (4.3) and (4.4) there is no term as *sine/cosine*, this reduces the memory space for storage of the values of *sine/cosine* completely. On the other hand, the simple values $\{1, 0, -1\}$ of Möbius function reduce the operations significantly, such that a great number of multiplicative operations in traditional FFT are replaced by additions and subtractions, and the latter is available to parallel algorithm. Also, the problem related to nonzero direct component will be solved in Bruns algorithm.

4.3 The Improvement of Wintner Algorithm by Reed

4.3.1 *Two other modified Möbius inverse formulas*

For convenience, we introduce two other theorems first.

Theorem 4.2.

$$G(x) = \sum_{n=1}^{[x]} F(\frac{x}{n}) \overset{x \geq 1}{\Leftrightarrow} F(x) = \sum_{n=1}^{[x]} \mu(n) G(\frac{x}{n}) \qquad (4.20)$$

where $[x]$ denotes the integer part of the real number x.

Fig. 4.3 Aurel Wintner (1903–1958)

Tufts Reed

Fig. 4.4 D.W.Tufts (1933–) and I.S.Reed (1923–)

Proof. In fact,
$$\sum_{n=1}^{[x]} \mu(n) G(\frac{x}{n}) = \sum_{n=1}^{[x]} \mu(n) \sum_{m=1}^{[x/n]} F(\frac{x}{mn})$$
$$= \sum_{k=1}^{[x]} F(\frac{x}{k}) \{\sum_{n|k} \mu(n)\} = F(x).$$

Here $mn = k$ is equivalent to $n|k$. Similarly, we have
$$F(x) = \sum_{n=1}^{[x]} \mu(n) G(\frac{x}{n}) \overset{x \geq 1}{\Rightarrow} G(x) = \sum_{n=1}^{[x]} F(\frac{x}{n}).$$
□

Theorem 4.3.
$$G(n) = \sum_{k=1}^{[N/n]} F(kn) \Leftrightarrow F(n) = \sum_{k=1}^{[N/n]} \mu(k) G(kn) \tag{4.21}$$

Proof. Let
$$R(n) = \sum_{k=1}^{[N/n]} \mu(k) G(kn)$$
then
$$R(n) = \sum_{k=1}^{[N/n]} \mu(k) \sum_{m=1}^{[N/(kn)]} F(mkn).$$
Denote $q = mk$, then
$$1 \leq m \leq [N/(kn)] \Rightarrow 1 \leq q \leq [N/(kn)]k \leq [N/n]$$
thus
$$R(n) = \sum_{q=1}^{[N/n]} F(qn) \sum_{mk=q} \mu(k) = \sum_{q=1}^{[N/n]} F(qn) \{\sum_{k|q} \mu(k)\} = F(n).$$
□

4.3.2 Reed's expression

In 1990, Reed proposed an improvement for the Wintner algorithm. He truncated the sum in Eqs. (4.2) and (4.3)[Ree90], thus

$$f(x) = \sum_{n=1}^{N}[a(n)\cos nx + b(n)\sin nx], \quad x \in (-\infty, \infty) \quad (4.22)$$

it implicates

$$a(n) = b(n) = 0, \quad \text{when} \quad n > N.$$

Corresponding to two infinite summations in Eqs. (4.2) and (4.3), there are two finite summations, i.e.,

$$a(n) = \sum_{m=1}^{[N/n]} \frac{\mu(m)}{mn} \sum_{s=1}^{mn} f(2\pi \frac{s}{mn}) \quad (4.23)$$

and

$$b(n) = (-1)^k \sum_{m=1}^{[N/n]} \frac{\mu(m)}{mn} \sum_{s=1}^{mn} f(2\pi \frac{s}{mn} + 2\pi \frac{1}{2^{q+2}}) \quad (4.24)$$

where $n = 1, 2, \cdots, N$, while k, q was defined as before

$$2^q(2k+1) = n, k, q \in \{0, 1, 2, ...\} \quad (4.25)$$

Here, we ignore the proof of Reed's algorithm. In fact, Reed *et al.* have shown that the above result is rigorous. Compared to traditional DFT and FFT, the multiplicative operations become very simple multiplications to reciprocal integers. Since the value of Möbius function only takes $1, 0, -1$, the remained calculations in AFT are only additions and subtractions. Note that even if we only consider the even components or cosine components, the number of sampling is already over the number of Fourier coefficients. If we want to calculate the sine components too, the number of samplings will increase significantly.

For example with $N = 4$, based on Eq. (4.23) it follows that

$$\begin{aligned}a(1) &= \sum_{m=1}^{4} \frac{\mu(m)}{m} \sum_{s=1}^{m} f(2\pi \frac{s}{m}) \\ &= \frac{\mu(1)}{1} f(2\pi \frac{1}{1}) + \frac{\mu(2)}{2} [f(2\pi \frac{1}{2}) + f(2\pi \frac{2}{2})] \\ &\quad + \frac{\mu(3)}{3} [f(2\pi \frac{1}{3}) + f(2\pi \frac{2}{3}) + f(2\pi \frac{3}{3})] \\ &\quad + \frac{\mu(4)}{4} [f(2\pi \frac{1}{4}) + f(2\pi \frac{2}{4}) + f(2\pi \frac{3}{4}) + f(2\pi \frac{4}{4})]\end{aligned} \quad (4.26)$$

$$a(2) = \sum_{m=1}^{2} \frac{\mu(m)}{2m} \sum_{s=1}^{2m} f(2\pi \frac{s}{2m}) = \frac{\mu(1)}{2}[f(2\pi \frac{1}{2}) + f(2\pi \frac{2}{2})]$$

$$+ \frac{\mu(2)}{4}[f(2\pi \frac{1}{4}) + f(2\pi \frac{2}{4}) + f(2\pi \frac{3}{4}) + f(2\pi \frac{4}{4})] \qquad (4.27)$$

$$a(3) = \frac{\mu(1)}{3} \sum_{s=1}^{3} f(2\pi \frac{s}{3})$$

$$= \frac{\mu(1)}{3}[f(2\pi \frac{1}{3}) + f(2\pi \frac{2}{3}) + f(2\pi \frac{3}{3})] \qquad (4.28)$$

$$a(4) = \frac{\mu(1)}{4} \sum_{s=1}^{4} f(2\pi \frac{s}{4})$$

$$= \frac{\mu(1)}{4}[f(2\pi \frac{1}{4}) + f(2\pi \frac{2}{4}) + f(2\pi \frac{3}{4}) + f(2\pi \frac{4}{4})] \qquad (4.29)$$

From this example, the advantage of AFT in reducing the number of multiplicative operations is clearly shown. A large number of complicated multiplications is replaced mainly by additions, subtractions as well as a few simple reciprocals of integers. Of course, there exists another series of problems attributed to nonuniform sampling. In this example with $N = 4$, in order to calculate a(1), a(2), a(3) and a(4), we need the following data

$$f(2\pi \frac{1}{4}), f(2\pi \frac{2}{4}), f(2\pi \frac{3}{4}), f(2\pi \frac{4}{4}), f(2\pi \frac{1}{3}), f(2\pi \frac{2}{3}).$$

This nonuniform sequence can also be expressed as

$$1/4,\ 1/3,\ 1/2,\ 2/3,\ 3/4,\ 1.$$

For uniform sampling, a variety of interpolations were proposed, but it leads to over-sampling. For example, corresponding to the above Farey sequence the uniform sequence should be

$$\frac{1}{12},\ \frac{1}{6},\ \frac{1}{4},\ \frac{1}{3},\ \frac{5}{12},\ \frac{1}{2},\ \frac{7}{12},\ \frac{2}{3},\ \frac{3}{4},\ \frac{5}{6},\ 1.$$

The sampling number may increase further if the odd components are involved based on Eq. (4.24).

$$b(1) = \sum_{m=1}^{4} \frac{\mu(m)}{m} \sum_{s=1}^{m} f(2\pi \frac{s}{m} + 2\pi \frac{1}{2^2})$$

$$= \frac{\mu(1)}{1} f(2\pi(\frac{1}{1}+\frac{1}{4})) + \frac{\mu(2)}{2}[f(2\pi(\frac{1}{2}+\frac{1}{4})) + f(2\pi(\frac{2}{2}+\frac{1}{4}))]$$

$$+ \frac{\mu(3)}{3}[f(2\pi(\frac{1}{3}+\frac{1}{4})) + f(2\pi(\frac{2}{3}+\frac{1}{4})) + f(2\pi(\frac{3}{3}+\frac{1}{4}))]$$

$$+ \frac{\mu(4)}{4}[f(2\pi(\frac{1}{4}+\frac{1}{4})) + f(2\pi(\frac{2}{4}+\frac{1}{4}))$$

$$+ f(2\pi(\frac{3}{4}+\frac{1}{4})) + f(2\pi(\frac{4}{4}+\frac{1}{4}))] \qquad (4.30)$$

$$b(2) = \sum_{m=1}^{2} \frac{\mu(m)}{mn} \sum_{s=1}^{mn} f(2\pi(\frac{s}{mn}+\frac{1}{8}))$$

$$= \frac{\mu(1)}{2}[f(2\pi(\frac{1}{2}+\frac{1}{8})) + f(2\pi(\frac{2}{2}+\frac{1}{8}))] + \frac{\mu(2)}{4}[f(2\pi(\frac{1}{4}+\frac{1}{8}))$$

$$+ f(2\pi(\frac{2}{4}+\frac{1}{8})) + f(2\pi(\frac{3}{4}+\frac{1}{8})) + f(2\pi(\frac{4}{4}+\frac{1}{8}))] \qquad (4.31)$$

$$b(3) = \frac{\mu(1)}{3} \sum_{s=1}^{3} f(2\pi(\frac{s}{3}+\frac{1}{4}))$$

$$= \frac{\mu(1)}{3}[f(2\pi(\frac{1}{3}+\frac{1}{4})) + f(2\pi(\frac{2}{3}+\frac{1}{4})) + f(2\pi(\frac{3}{3}+\frac{1}{4}))] \qquad (4.32)$$

$$b(4) = \frac{\mu(1)}{4} \sum_{s=1}^{4} f(2\pi(\frac{s}{4}+\frac{1}{16}))$$

$$= \frac{\mu(1)}{4}[f(2\pi(\frac{1}{4}+\frac{1}{16}))+$$

$$+ f(2\pi(\frac{2}{4}+\frac{1}{16})) + f(2\pi(\frac{3}{4}+\frac{1}{16})) + f(2\pi(\frac{4}{4}+\frac{1}{16}))] \qquad (4.33)$$

The corresponding sampling sequence is $\frac{5}{16}, \frac{3}{8}, \frac{9}{16}, \frac{7}{12}, \frac{5}{8}, \frac{3}{4}, \frac{13}{16}, \frac{7}{8}, \frac{11}{12}, 1, \frac{17}{16}, \frac{9}{8}, \frac{5}{4}$. If we want to get uniform sampling, then other 32 sampling points are needed. If we want to cover sine components $\{b_n\}$, then we need additional 11 points. This is a typical over-sampling problem.

4.4 Fundamental Theorem of AFT (Bruns)

Theorem 4.4. In Bruns' algorithm of AFT, the Fourier coefficients should be expressed as

$$a(0) = \frac{1}{2\pi} \int_0^{2\pi} f(x)dx \tag{4.34}$$

and

$$a(n) = \sum_{l=1,3,5,\cdots}^{[N/n]} \mu(l)B(2nl,0) \tag{4.35}$$

as well as

$$b(n) = \sum_{l=1,3,5,\cdots}^{[N/n]} \mu(l)(-1)^{(l-1)/2} B(2nl, \frac{1}{4nl}) \tag{4.36}$$

where we define the alternating average $B(2n,\alpha)$ of the periodic function $f(x)$ with phase shift as follows.

$$B(2n,\alpha) = \frac{1}{2n} \sum_{m=1}^{2n} (-1)^m f(2\pi \frac{m}{2n} + 2\pi\alpha) \tag{4.37}$$

where n is an integer.

In the following subsections we will show a proof for Eqs. (4.35) and (4.36). For convenience, we repeat the definition of $S(n,\alpha)$ in (4.16) as

$$S(n,\alpha) = \sum_{r=1}^{\infty} a_{nr} = \frac{1}{n} \sum_{s=1}^{n} f(2\pi \frac{s}{n} + 2\pi\alpha) \tag{4.38}$$

Fig. 4.5 Ernst Heinrich Bruns (1848–1919)

this expression of the n-value average has involved phase shift. The proof includes several steps as follows.

(i) First, we will prove

$$S(n,\alpha) = a(0) + \sum_{m=1}^{[N/n]} c(mn,\alpha) \tag{4.39}$$

Successively, from the Möbius inverse formula Eq. (4.21) it follows that

$$c(n,\alpha) = \sum_{m=1}^{[N/n]} \mu(m)[S(mn,\alpha) - a(0)]. \tag{4.40}$$

Note that if the upper limit is infinite, it is impossible to have this kind of formula except if the direct component is vanished.

(ii) Second, we will show that

$$B(2n,0) = \sum_{l=1,3,5,\ldots}^{[N/n]} a(ln) \tag{4.41}$$

and

$$B(2n,\frac{1}{4n}) = \sum_{l=1,3,5,\ldots}^{[N/n]} (-1)^{(\ell-1)/2} b(ln). \tag{4.42}$$

(iii) Next, based on the modified Möbius inverse formulas, Eqs. (4.35) and (4.36) can be obtained directly. In other words, the Bruns algorithm of *AFT* formulation is given.

4.4.1 Proof of Eq. (4.41)

$c(n,\alpha)$ and $d(n,\alpha)$ are defined as

$$c(n,\alpha) = a(n)\cos 2\pi n\alpha + b(n)\sin 2\pi n\alpha \tag{4.43}$$

and

$$d(n,\alpha) = -a(n,\alpha)\sin 2\pi n\alpha + b(n,\alpha)\cos 2\pi n\alpha \tag{4.44}$$

respectively, where $\alpha \in (-1,1)$. Therefore,

$$f(t+\alpha T) = a(0) + \sum_{k=1}^{N} c(k,\alpha)\cos 2\pi k f_0 t + \sum_{k=1}^{N} d(k,\alpha)\sin 2\pi k f_0 t \tag{4.45}$$

thus

$$S(n,\alpha) = \frac{1}{n}\sum_{m=1}^{n} f(\frac{m}{n}T + \alpha T)$$

$$= \frac{1}{n}\sum_{m=1}^{n}[a(0) + \sum_{k=1}^{N} c(k,\alpha)\cos(\frac{2\pi km}{n}) + \sum_{k=1}^{N} d(k,\alpha)\sin(\frac{2\pi km}{n})]$$

$$= a(0) + \sum_{k=1}^{N} c(k,\alpha)\frac{1}{n}\sum_{m=1}^{n}\cos(\frac{2\pi km}{n}) + \sum_{k=1}^{N} d(k,\alpha)\frac{1}{n}\sum_{m=1}^{n}\sin(\frac{2\pi km}{n})$$

$$= a(0) + \sum_{k=1}^{N} c(k,\alpha)\frac{1}{n}\sum_{m=1}^{n}\cos(\frac{2\pi km}{n}).$$

Therefore,

$$S(n,\alpha) = a(0) + \sum_{r=1}^{[N/n]} c(rn,\alpha) \qquad (4.46)$$

By using Möbius inverse formula, we have

$$c(n,\alpha) = \sum_{r=1}^{[N/n]} \mu(r)[S(rn,\alpha) - a(0)] \qquad (4.47)$$

The above derivation identifies Eqs. (4.7) and (4.8) again.

4.4.2 *The relationship between $a(n), b(n)$ and $B(2n,\alpha)$*

According to Eq. (4.37), the alternating average $B(2n,\alpha)$ of the periodic function $f(x)$ with phase shifts is

$$B(2n,\alpha) = \frac{1}{2n}\sum_{m=1}^{2n}(-1)^m f(\frac{m2\pi}{2n} + 2\pi\alpha)$$

where n is an integer. Now we are going to prove Eqs. (4.41) and (4.42). Since we have

$$B(2n,0) = \frac{1}{2n}\sum_{m=1}^{2n}(-1)^m f(\frac{m2\pi}{2n})$$

$$= \frac{1}{2n}[\sum_{m=1}^{n} f(\frac{m2\pi}{n}) - \sum_{m=1}^{n} f(\frac{m2\pi}{n} + \frac{2\pi}{2n})]$$

and
$$S(n,\alpha) = a(0) + \sum_{l=1}^{\infty} c(ln,\alpha)$$
$$= a(0) + [\sum_{l=1}^{\infty} (a(ln)\cos(2\pi ln\alpha) + b(ln)\sin(2\pi ln\alpha))].$$
Therefore, it is given by
$$B(2n,0) = \frac{1}{2}[S(n,0) - S(n,\frac{1}{n})]$$
$$= \frac{1}{2}\sum_{l=1}^{[N/n]} [a(ln) - a(ln)\cos l\pi] = \sum_{l=1,3,5,\ldots}^{[N/n]} a(ln). \tag{4.48}$$
This is the proof for Eq. (4.41). By using the Möbius inverse formula Eq. (4.21) with all the divisors are odd numbers, Eq. (4.35) is given directly by
$$a(l) = \sum_{l=1,3,5,\ldots}^{\infty} \mu(l)B(2ln,0).$$
Similarly, we have
$$B(2n,\frac{1}{4n}) = \frac{1}{2}\sum_{m=1}^{2n} f(\frac{m2\pi}{2n} + \frac{2\pi}{4n})(-1)^m$$
$$= \frac{1}{2}[\sum_{m=1}^{n} f(\frac{m2\pi}{n} + \frac{2\pi}{4n}) - \sum_{m=1}^{n} f(\frac{m2\pi}{n} + \frac{2\pi}{2n} + \frac{2\pi}{4n})]$$
$$= \frac{1}{2}[S(n,\frac{1}{4n}) - S(n,\frac{3}{4n})]$$
$$= \frac{1}{2}\sum_{l=1}^{\infty} [a(\ell n)\cos\frac{\ell\pi}{2} + b(\ell n)\sin\frac{\ell\pi}{2} - a(\ell n)\cos\frac{3\ell\pi}{2} - b(ln)\sin\frac{3\ell\pi}{2}]$$
$$= \sum_{l=1,3,5,\ldots}^{\infty} (-1)^{(\ell-1)/2} b(\ell n).$$
This is the proof for Eq. (4.42). Thus Eq. (4.36) is given by
$$b(n) = \sum_{l=1,3,5,\ldots}^{[N/n]} \mu(l)(-1)^{(l-1)/2} B(2nl,\frac{1}{4nl}).$$
In the above derivation, the other two theorems are used as follows.

Theorem 4.5.
$$g(n) = \sum_{k=1,3,5,\ldots}^{[N/n]} f(kn) \Rightarrow f(n) = \sum_{k=1,3,5,\ldots}^{[N/n]} \mu(k)g(kn), \tag{4.49}$$

and

Theorem 4.6.

$$g(n) = \sum_{k=1,3,5,\ldots}^{[N/n]} (-1)^{(k-1)/2} f(kn)$$

$$\Rightarrow f(n) = \sum_{m=1,3,5,\ldots}^{[N/n]} \mu(m)(-1)^{(m-1)/2} g(mn) \quad (4.50)$$

The proof of Eq. (4.49) is similar to that for Eq. (4.21) except that both k and m are odd (thus q is odd too). In order to prove Eq. (4.50), denoting $k = 2i - 1, m = 2j - 1$ and $q = 2s - 1$, and comparing

$$\frac{k-1}{2} + \frac{m-1}{2} = (i-1) + (j-1) = (i+j) - 2$$

and

$$q = 2s - 1 = (2i - 1)(2j - 1) = 4ij - 2(i+j) + 1$$

or

$$\frac{q-1}{2} = 2ij - (i+j)$$

it is concluded that

$$(-1)^{\frac{k-1}{2}}(-1)^{\frac{m-1}{2}} = (-1)^{\frac{q-1}{2}}.$$

Note that

$$[([N/n]+1)/2] = \begin{cases} [N/n]/2 & \text{when } [N/n] \text{ is even} \\ ([N/n]+1)/2 & \text{when } [N/n] \text{ is odd} \end{cases}.$$

Thus the summation relation is given by

$$\sum_{k=1,3,5,\cdots}^{[N/n]} (-1)^{\frac{k-1}{2}} \Rightarrow \sum_{i=1}^{[([N/n]+1)/2]} (-1)^{i-1}$$

such as

$$g(n) = \sum_{i=1}^{[([N/n]+1)/2]} (-1)^{i-1} f((2i-1)n) \Rightarrow$$

$$f(n) = \sum_{j=1}^{[([N/n]+1)/2]} (-1)^{i-1} \mu((2j-1))(-1)^{j-1} g((2j-1)n) \quad (4.51)$$

Fig. 4.6 A postage stamp of Srinivasa Ramanujan (1887–1920)

4.5 Uniformly Sampling in AFT based on Ramanujan Sum

This section presents the Ramanujan sum rule in order to improve the AFT.

4.5.1 *What is the Ramanujan sum rule?*

The definition of Ramanujan sum is

$$C(m,n) = \sum_{\substack{h \in [1,n] \\ (h,n)=1}} e^{2\pi i h m/n} \qquad (4.52)$$

in which h runs over only the integer values less than n and prime to it. This is an unusual triangular summation. There is an interesting theorem related to the Ramanujan sum as follows.

Theorem 4.7.

$$\sum_{\substack{h \in [1,n] \\ (h,n)=1}} e^{2\pi i h m/n} = \sum_{d|(m,n)} d\mu(n/d) \qquad (4.53)$$

where d runs over the common divisors of m and n. From the right hand side of (4.53), Ramanujun sum $C(m,n)$ must be an integer. Note that taking $m=1$, we may obtain another definition of Möbius function

Definition 4.1.

$$\mu(n) = \sum_{\substack{1 \leq h \leq n \\ (h,n)=1}} e^{2\pi i h/n} \qquad (4.54)$$

Surely, one is bound to ask, there must be some physical meaning in it? Before answering this, we are going to prove the Ramanujan sum rule.

4.5.2 Proof of Ramanujan sum rule

Considering n rational fractions

$$\frac{h}{n}, \quad h \in [1, n] \tag{4.55}$$

in which each fraction can be expressed as a reducible one uniquely as

$$\frac{h}{n} = \frac{a}{d} \tag{4.56}$$

where a and d are uniquely determined by h and n such that $d \mid n$ and

$$(a, d) = 1 \quad a \in [1, d]. \tag{4.57}$$

Conversely, every fraction a/d with $d \mid n$ and $(a, d) = 1$ for $a \in [1, d]$ occurs in the set h/n with $h \in [1, n]$, although generally not in reduced form. Therefore, for any function F(x) it is obvious that

$$\sum_{1 \leq h \leq n} F(\frac{h}{n}) = \sum_{d \mid n} \sum_{\substack{a \in [1, d] \\ (a, d) = 1}} F(\frac{a}{d}). \tag{4.58}$$

Now let

$$g(n) = \sum_{1 \leq h \leq n} F(\frac{h}{n}) \tag{4.59}$$

and

$$f(n) = \sum_{\substack{h \in [1, n] \\ (h, n) = 1}} F(\frac{h}{n}) \tag{4.60}$$

then Eq. (4.58) becomes

$$g(n) = \sum_{d \mid n} f(d). \tag{4.61}$$

According to the classical Möbius inverse formula, we have

$$f(n) = \sum_{d \mid n} \mu(\frac{n}{d}) g(d) \tag{4.62}$$

or

$$\sum_{\substack{h \in [1, n] \\ (h, n) = 1}} F(\frac{h}{n}) = \sum_{d \mid n} \mu(\frac{n}{d}) \sum_{1 \leq a \leq d} F(\frac{a}{d}). \tag{4.63}$$

Take $F(x) = e^{imx}$, then we have

$$f(n) = \sum_{\substack{1 \leq h \leq n \\ (h,n) = 1}} e^{2\pi i \frac{mh}{n}} \equiv C(m,n) \tag{4.64}$$

and

$$g(n) = \sum_{1 \leq h \leq n} e^{2\pi i \frac{mh}{n}} = \begin{cases} n, & if \ n|m \\ 0, & if \ n \nmid m \end{cases} \tag{4.65}$$

Therefore,

$$C(m,n) = f(n) = \sum_{d|n \atop d|m} \mu(\frac{n}{d}) g(d) = \sum_{d|(m,n)} \mu(\frac{n}{d}) d \tag{4.66}$$

Note that in the Ramanujan sum rule Eq. (4.53), let $m = 1$, then

$$\mu(n) = \sum_{\substack{1 \leq h \leq n \\ (h,n) = 1}} e^{2\pi i h/n} \tag{4.67}$$

This can be considered as another definition of Möbius function.

4.5.3 Uniformly sampling AFT (USAFT)

Now let us discuss an important application of Ramanujan sum as follows. It is well known that in Tufts–Reed algorithm Eqs. (4.22)–(4.26), let us consider the case of N being a multiple of 4. For $n|N$, we have

$$a_n(N) = \sum_{m|\frac{N}{n}} \frac{\mu(n)}{mn} \sum_{s=1}^{mn} f(2\pi \frac{s}{mn})$$

$$= \sum_{d|\frac{N}{n}} \frac{\mu(\frac{N/n}{d})}{\frac{N/n}{d} n} \sum_{s=1}^{\frac{N/n}{d} n} f(2\pi \frac{s}{\frac{N/n}{d} n})$$

$$= \sum_{r=1}^{N} \sum_{d|r \atop d|\frac{N}{n}} \frac{d}{N} \mu(\frac{N}{nd}) f(2\pi \frac{r}{N})$$

$$= \frac{1}{N} \sum_{r=1}^{N} \sum_{d|(r,\frac{N}{n})} d\mu(\frac{N}{nd}) f(2\pi \frac{r}{N}) \tag{4.68}$$

Therefore, the Fourier expansion coefficients $a_n(N)$ can be obtained by sampled values of function $f(x)$ under consideration with weights of Ramanujan's sum as

$$a_n(N) = \frac{1}{N}\sum_{r=1}^{N} C(r, \frac{N}{n})f(2\pi\frac{r}{N}). \quad (4.69)$$

Similarly, when $n|\frac{N}{4}$ we have

$$b_n(N) = (-1)^k \sum_{m|\frac{N}{n}} \frac{\mu(m)}{mn} \sum_{s=1}^{mn} f(2\pi\frac{s}{mn} + 2\pi\frac{1}{2^{q+2}})$$

$$= \frac{(-1)^k}{N} \sum_{r=1}^{N} C(r, \frac{N}{n})f(2\pi(\frac{r}{N} + \frac{1}{2^q}))$$

$$= \frac{(-1)^k}{N} \sum_{r=1}^{N} C(r - \frac{N}{2^{q+2}}, \frac{N}{n})f(2\pi\frac{r}{N}). \quad (4.70)$$

This theorem shows that the coefficients of USAFT simply equal to Ramanujan's sum. By Eq. (4.53), each Ramanujan's sum can be evaluated by addition and subtraction of some integers. Therefore, we conclude that USAFT algorithm can be evaluated using parallel processing with uniform sampling.

4.5.3.1 Example for $N = 4$

Four sampling points are required in this case, that is, $f(0), f(\frac{\pi}{2}), f(\pi)$ and $f(\frac{3\pi}{2})$. Correspondingly, we have

$$\begin{pmatrix} a_1 \\ a_2 \\ a_4 \\ b_1 \end{pmatrix} = \frac{1}{4} \begin{pmatrix} C(1,4) & C(2,4) & C(3,4) & C(4,4) \\ C(1,2) & C(2,2) & C(3,2) & C(4,2) \\ C(1,1) & C(2,1) & C(3,1) & C(4,1) \\ C(0,4) & C(1,4) & C(2,4) & C(3,4) \end{pmatrix} \begin{pmatrix} f(\frac{1}{4}2\pi) \\ f(\frac{2}{4}2\pi) \\ f(\frac{3}{4}2\pi) \\ f(\frac{4}{4}2\pi) \end{pmatrix}$$

$$= \frac{1}{4} \begin{pmatrix} 0 & -2 & 0 & 2 \\ -1 & 1 & -1 & 1 \\ 1 & 1 & 1 & 1 \\ 2 & 0 & -2 & 0 \end{pmatrix} \begin{pmatrix} f(\frac{1}{4}2\pi) \\ f(\frac{2}{4}2\pi) \\ f(\frac{3}{4}2\pi) \\ f(\frac{4}{4}2\pi) \end{pmatrix} \quad (4.71)$$

Note that all the coefficients $C(m, n)$ can be calculated by using Ramanujan sum rule, that is

$$C(m, n) = \sum_{d|(m,n)} d \cdot \mu(\frac{n}{d})$$

For example, we have

$$C(3, 4) = \sum_{d|(3,4)} d \cdot \mu(4/d) = \mu(4) = 0.$$

4.5.3.2 Example for $N = 8$

In this case, 8 values of $f(x)$ are needed, that is, $\{f(\frac{s}{8}2\pi)\}(s = 1, 2, ..., 8)$. Note that $C(m, n) = C(m + n, n)$, then the corresponding result is as follows.

$$\begin{pmatrix} a_1 \\ a_2 \\ a_4 \\ a_8 \\ b_1 \\ b_2 \end{pmatrix} = \frac{1}{8} \begin{pmatrix} C(1,8)C(2,8)C(3,8)C(4,8)C(5,8)C(6,8)C(7,8)C(8,8) \\ C(1,4)C(2,4)C(3,4)C(4,4)C(5,4)C(6,4)C(7,4)C(8,4) \\ C(1,2)C(2,2)C(3,2)C(4,2)C(5,2)C(6,2)C(7,2)C(8,2) \\ C(1,1)C(2,1)C(3,1)C(4,1)C(5,1)C(6,1)C(7,1)C(8,1) \\ C(7,8)C(0,8)C(1,8)C(2,8)C(3,8)C(4,8)C(5,8)C(6,8) \\ C(0,4)C(1,4)C(2,4)C(3,4)C(4,4)C(5,4)C(6,4)C(7,4) \end{pmatrix} \begin{pmatrix} f(\frac{1}{8}2\pi) \\ f(\frac{2}{8}2\pi) \\ f(\frac{3}{8}2\pi) \\ f(\frac{4}{8}2\pi) \\ f(\frac{5}{8}2\pi) \\ f(\frac{6}{8}2\pi) \\ f(\frac{7}{8}2\pi) \\ f(\frac{8}{8}2\pi) \end{pmatrix}$$

$$= \frac{1}{8} \begin{pmatrix} 0 & 0 & 0 & -4 & 0 & 0 & 0 & 4 \\ 0 & -2 & 0 & 2 & 0 & -2 & 0 & 2 \\ -1 & 1 & -1 & 1 & -1 & 1 & -1 & 1 \\ 1 & 1 & 1 & 1 & 1 & 1 & 1 & 1 \\ 0 & 4 & 0 & 0 & 0 & 4 & 0 & 0 \\ 2 & 0 & -2 & 0 & 2 & 0 & -2 & 0 \end{pmatrix} \begin{pmatrix} f(\frac{1}{8}2\pi) \\ f(\frac{2}{8}2\pi) \\ f(\frac{3}{8}2\pi) \\ f(\frac{4}{8}2\pi) \\ f(\frac{5}{8}2\pi) \\ f(\frac{6}{8}2\pi) \\ f(\frac{7}{8}2\pi) \\ f(\frac{8}{8}2\pi) \end{pmatrix} \quad (4.72)$$

Note that in the above equation a non-square 6×8 matrix is used. All the matrix elements can be evaluated by Eq. (4.53) using only additions and simple multiplications. At first sight, this apparently gives us a hope of uniformly sampled AFT. But the uniform sampling in sample space $\{f(\frac{2\pi s}{N})\}$ does not correspond to a uniform distribution in Fourier space, that is $a_1, a_2, a_4, \ldots, a_n, \ldots$, with $n \mid N$. The next subsection demonstrates the method to obtain the Fourier coefficient corresponding to the first t integers.

4.5.3.3 $N = 4[1, 2, 3, ..., t]$

If N is equal to 4 multiplied by the minimum common multiple of the first t natural numbers, then from Eqs. (4.69) and (4.70) the Fourier coefficients $a_1, a_2, a_3, \cdots, a_t, \cdots$ can be evaluated. It is shown that for the uniform sampling in AFT, we have to face a serious over-sampling problem as before. However, this technique sufficiently reveals the duality of arithmetic Fourier transform. On the one hand, the uniform distribution in Fourier space leads to non-uniformly distribution in sampling space; on the other hand, the uniform sampling leads to nonuniform distribution in Fourier space. For any particular calculation requiring uniform sampling, we have a serious problem of over-sampling. In order to overcome this problem, we have to adjust proportions for both spaces.

4.5.4 Note on application of generalized function

For convenience let us consider the case of a periodic even function $F(x)$ with vanished direct component.

$$a_0 = \frac{1}{\pi} \int_0^{2\pi} F(x)dx = 0$$

Then

$$a_n = \frac{1}{\pi} \int_0^{\infty} F(x) \cos nx \, dx \qquad (4.73)$$

In the theory of generalized functions, there is a well-known theorem that

Theorem 4.8.

$$\sum_{m=1}^{\infty} \cos mx = -\frac{1}{2} + \pi \sum_{m=-\infty}^{\infty} \delta(x - 2m\pi) \qquad (4.74)$$

or

$$\sum_{m=1}^{\infty} \cos mnx = -\frac{1}{2} + \pi \sum_{m=-\infty}^{\infty} \delta(nx - 2m\pi) \qquad (4.75)$$

Therefore,

$$\sum_{m=1}^{\infty} a_{nm} = \sum_{m=1}^{\infty} \frac{1}{\pi} \int_0^{2\pi} F(x) \cos nm \, x \, dx$$

$$= \frac{1}{\pi} \int_0^{2\pi} F(x) \{ \sum_{m=1}^{\infty} \cos nmx \} dx$$

$$= \frac{1}{\pi} \int_0^{2\pi} F(x) \{ \frac{1}{2} + \pi \sum_{m=-\infty}^{\infty} \delta(nx - 2m\pi) \, dx$$

$$= \sum_{-m=\infty}^{\infty} \int_0^{2\pi} F(x) \delta(nx - 2m\pi) \, dx$$

$$= \sum_{m=-\infty}^{\infty} \int_0^{2\pi} F(x) \delta(nx - 2m\pi) d\frac{(nx - 2m\pi)}{n} \qquad (4.76)$$

Let $y = nx - 2m\pi$, then
$$\sum_{m=1}^{\infty} a_{nm} = \frac{1}{n} \sum_{m=-\infty}^{\infty} \int_{-2m\pi}^{2n\pi - 2m\pi} F(\frac{y + 2m\pi}{n})\delta(y) \, dy$$
$$= \frac{1}{n} \sum_{m=1}^{n} \int_{-2m\pi}^{2n\pi - 2m\pi} F(\frac{y + 2m\pi}{n})\delta(y) \, dy$$
$$= \frac{1}{n} \sum_{m=1}^{n} F(\frac{2m\pi}{n}) = \frac{1}{n} \sum_{m=1}^{n} F(-\frac{2m\pi}{n}) \equiv S_n \qquad (4.77)$$

Therefore, it follows that
$$\sum_{m=1}^{\infty} a_{mn} = S(n) \qquad (4.78)$$

Again, we use the Möbius inverse formula such that
$$a_n = \sum_{m=1}^{\infty} \mu(m) S(mn) = \sum_{m=1}^{\infty} \frac{\mu(m)}{mn} \sum_{s=1}^{mn} F(\frac{2\pi s}{mn}) \qquad (4.79)$$

This result is just the same as Eq. (4.2). Considering the relation between cosine functions and Dirac $\delta-$ function, it can be understood why the harmonic function *sine/cosine* disappears in AFT algorithm.

4.6 Summary

Chapter 4 presented the application of a variety of Möbius inversion formulas to arithmetic Fourier transform (AFT), which is crucial to parallel processing, and also introduced different versions of AFT:

(1) Wintner's theorem (for only cosine terms) and Reed's algorithm.

(2) Bruns' theorem (for both cosine and sine) and Tufts' algorithm.

(3) For overcoming the nonuniform sampling existing in all previous work, a uniform sampling AFT (USAFT) algorithm was addressed based on the Ramanujan sum rule. A comparison between Tufts' AFT and USAFT is shown as follows.

(i)Expression of $f(x)$ in Tufts AFT:
$$f(x) = \sum_{n=1}^{N} a_n \cos nx + \sum_{n=1}^{N} b_n \sin nx$$

Expression of $f(x)$ in USAFT:
$$f(x) = \sum_{n|N} a_n \cos nx + \sum_{n|\frac{N}{4}} b_n \sin nx$$

(ii) Sum rule in Tufts' AFT:
$$\frac{1}{N}\sum_{s=1}^{N} e^{i2\pi s \frac{m}{N}} = \begin{cases} 1, N \mid m \\ 0, N \nmid m \end{cases}$$

Sum rule in USAFT:
$$\frac{1}{N}\sum_{s=1}^{N} C(s, \frac{N}{n}) e^{i2\pi s \frac{m}{N}} = \begin{cases} 1, m = n \\ 0, m \neq n, m \mid N \end{cases}$$

(iii) The expansion coefficients in Tufts. AFT:
$$a_n = \sum_{m=1}^{N/n} \frac{\mu(m)}{mn} \sum_{s=1}^{mn} f(\frac{2\pi s}{mn})$$

The expansion coefficients in USAFT:
$$a_n = \frac{1}{N}\sum_{n \mid N} C(s, \frac{N}{n}) f(\frac{2\pi s}{N})$$

Obviously, the AFT algorithm only contains additions and subtractions, this is suitable for parallel processing, and the USAFT has gone further more. For example, a rectangular matrix relation is given such as Eqs. (4.71) and (4.72).

Note that the the other application of the sum rule in USAFT is left for future study. Also, the details of the Reed algorithm, arithmetic Z-transform and the convolution in AFT have not been discussed in this chapter. Two dimensional AFT will be introduced in Chapter 5.

Chapter 5

Inverse Lattice Problems in Low Dimensions

5.1 Concept of Low Dimensional Structures

From the mathematical point of view, the one-dimensional (1D) atomic chain is a very easily described periodic structure. Regarding periodicity, the congruence analysis arose thousands of years ago, the Fourier analysis was initiated 200 years ago, and the group theory, as well as reciprocal analysis, appeared almost 100 years ago. However, in material science the linear chain was only a simplified model for understanding a three dimensional crystal structure in pedagogy. Also, only 50 years before, it was well recognized in physics society that "No long range order in 2D, dislocation should appear at any finite temperature. And similar results hold in one dimension"[Mer68].

Very recently Ohnishi et al., using a combination of transmission electron microscopy(TEM) and scanning tunnelling microscopy, visualized the formation of monatomic gold wires[Ohn98]. Also, the appearance of graphene(Fig. 5.1) (a single graphite sheet with ripples) has made the research of monatomic layer film a popular subject. A variety of nanowires and monolayer films often exhibit novel properties which are both interesting in low dimensional physics and technically important for application to nano-devices[1]. Correspondingly, many classical mathematical theorems, only having pure theoretical meaning in the old days, become realistic and applicable now. As G.C. Rota said, the mystery of mathematics is that conclusions originating in the play of the mind do find striking practical applications[Dav81].

This chapter mainly discusses the inverse lattice problems in low dimensions by using the Möbius inversion method.

---
[1]Note that they are metastable.

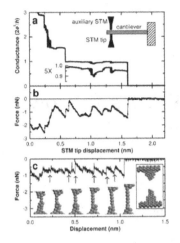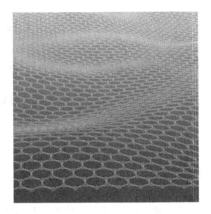

Fig. 5.1 Experiment on Au monatomic chain and an image of a Graphene sheet

5.2 Linear Atomic Chains

In Chapter 1, the inversion formula for a monatomic chain is given as

$$E(x) = \sum_{n=1}^{\infty} \Phi(nx) \Leftrightarrow \Phi(x) = \sum_{n=1}^{\infty} \mu(n) E(nx) \tag{5.1}$$

The theoretical investigations of atomic chains have received a lot of attention in the last decade, since single gold chains were produced between two gold electrodes[Ohn98; Yan98]. The researches focus on the stabilities of linear chains, zigzag chains, ladders based on various metal atoms Au, Al, Cu, Pd, Pt, Ru, Rh, Ag, K, Ga. Concretely, based on first principle calculations for the band structures and electronic density distributions of these metallic structures, one obtains the bond length and bond angle of the stable and meta-stable structures. Figures 5.2 and 5.3 show the *ab initio* calculated cohesive energies for linear atomic chains of Au, Al, Ga and corresponding converted interatomic potentials in these chains. Note that there are two or more minima in each cohesive energy curve either for Al or Ga, in which p-electrons play an important role. On the other hand, there exists a Peierls phase transition from linear chain to zigzag chain. For Cu and Au wires the zigzag distortion is favorable even when the linear wire is stretched, but it is not observed for K and Ca wire[Por2001]. Most stable structures with zigzag are close to equilateral triangle with deviation around 60° and 120° since distortion of high symmetry quite often leads to increasing cohesive energy.

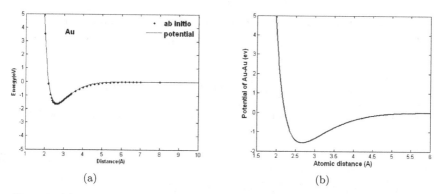

Fig. 5.2 (a)Cohesion curve of Au chain. (b)Interatomic potentials in Au chain.

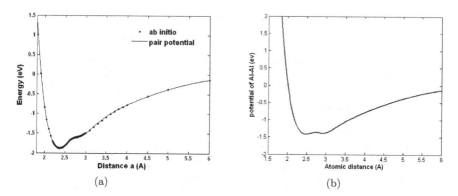

Fig. 5.3 (a)Cohesion curve of Al chain. (b)Interatomic potentials in Al chain.

5.3 Simple Example in a Square Lattice

Now let us consider a 2D square lattice. The cohesive energy can be expressed as

$$\mathbb{E}(x) = \frac{1}{2} \sum_{n=1}^{\infty} r(n) \Phi(b(n)x) \tag{5.2}$$

The concrete data for the coordination number $r(n)$ and distance function $b(n)$ are shown in Table 5.1. Note that b(n) can be expressed as $|p + iq|$, and

$$|p_1 + iq_1| \cdot |p_2 + iq_2| = |(p_1 p_2 - q_1 q_2) + i(p_1 q_2 + q_1 p_2)|$$

this means that for any integer pair (m,n) there must be an integer k such that

$$b(m)b(n) = b(k) \tag{5.3}$$

In other words, $\{b(n)\}$ is a multiplicative semigroup.

Here let us define $I(n)$ by a new sum rule, that is, a recursive relation

$$\sum_{b(m)b(n)=b(k)} I(n)r(m) = \delta_{k,1} \tag{5.4}$$

Now the Möbius inverse theorem for 2D square lattice is given by

Table 5.1 Distance function $b(n)$ and coordination function $r(n)$ in 2D square lattice

n	1	2	3	4	5	6	7	8	9	10
b(n)	1	$\sqrt{2}$	2	$\sqrt{5}$	$\sqrt{8}$	3	$\sqrt{10}$	$\sqrt{13}$	4	$\sqrt{17}$
r(n)	4	4	4	8	4	4	8	8	4	8
n	11	12	13	14	15	16	17	18	19	20
b(n)	$\sqrt{18}$	$\sqrt{20}$	5	$\sqrt{26}$	$\sqrt{29}$	$\sqrt{32}$	$\sqrt{34}$	6	$\sqrt{37}$	$\sqrt{40}$
r(n)	4	8	8+4	8	8	4	8	4	8	8
n	21	22	23	24	25	26	27	28	29	30
b(n)	$\sqrt{41}$	$\sqrt{45}$	7	$\sqrt{50}$	$\sqrt{52}$	$\sqrt{53}$	$\sqrt{58}$	$\sqrt{61}$	8	$\sqrt{65}$
r(n)	8	8	4	8+4	8	8	8	8	4	8+8

Theorem 5.1.

$$\mathbb{E}(x) = \frac{1}{2}\sum_{n=1}^{\infty} r(n)\Phi(b(n)x)$$

$$\xrightarrow[(5.3)]{(5.4)}$$

$$\Phi(x) = 2\sum_{n=1}^{\infty} I(n)\mathbb{E}(b(n)x). \tag{5.5}$$

Proof.

$$2\sum_{n=1}^{\infty} I(n)\mathbb{E}(b(n)x)$$

$$= 2\sum_{n=1}^{\infty} I(n)\frac{1}{2}\sum_{m=1}^{\infty} r(m)\mathbb{E}(b(m)[b(n)x])$$

$$= \sum_{k=1}^{\infty}\left\{\sum_{b(m)b(n)=b(k)} I(n)r(m)\right\}\mathbb{E}(b(k)x)$$

$$= \sum_{k=1}^{\infty} \delta_{k,1}\mathbb{E}(b(k)x) = \mathbb{E}(b(1)x) = \mathbb{E}(x).$$

□

Obviously, the reasoning of Eq. (5.5) is only determined by closeness of $\{b(n)\}$ regarding to multiplicative operations. Therefore, there must exist the corresponding Möbius inverse formula provided that a lattice has the multiplicative property. Note that Eq. (5.4) can also be expressed as

$$\sum_{b(n)|b(k)} I(n)r(b^{-1}\{\frac{b(k)}{b(n)}\}) = \delta_{k,1} \tag{5.6}$$

where $b(n)|b(k)$ means that b(n) runs over all the factors of b(k), including b(1) and b(k). Here the concept of factor means $\frac{b(k)}{b(n)} \in \{b(n)\}$. The operator of b^{-1} is defined as $b^{-1}[b(n)] = n$.

It can be concluded that there must exist a Möbius inverse formula for any lattice, provided that the multiplicative closure is held, whether the concrete lattice is 2- or 3-dimensional. This conclusion will be used in Chapter 6 in order to establish a general Möbius inverse formula in an arbitrary lattice. Also, the concept of closure regarding multiplication will be applied in succeeding sections in this chapter but the definition of multiplication will be little different.

5.4 Arithmetic Functions on Gaussian Integers

5.4.1 Gaussian integers

A two-dimensional square lattice can be represented by the whole set of complex integers $m + in$, or the Gaussian integers $G = m + in$. Note that if $\alpha, \beta \in G$, then $\alpha \pm \beta$ and $\alpha\beta \in G$. Also, regarding the addition, for any element $\alpha \in G$ there exists a zero element 0 and a negative element β such that $\alpha + 0 = \alpha$ and $\alpha + \beta = 0$. Simultaneously, for any element $\alpha \in G$, regarding the multiplication there is a unit element '1' such that $\alpha \cdot 1 = 1 \cdot \alpha = \alpha$, and there is also distributive property of multiplication regarding addition, i.e., $\alpha(\beta + \gamma) = \alpha\beta + \alpha\gamma$[Pan2001; Nar89]. In this chapter, we focus on the Gaussian integers $\widetilde{G}$ with $\alpha = m + in$ and $(m,n) \neq (0,0)$. $\widetilde{G}$ is a semigroup regarding multiplicative operation.

Definition 5.1. Any function on $\widetilde{G}$ is called an arithmetic function.

In order to discuss this kind of function, we have to become familiar with the multiplicative relationship among Gaussian integers in $\widetilde{G}$.

5.4.2 Unit elements, associates with reducible and irreducible integers in $\widetilde{G}$

Definition 5.2. A Gaussian integer γ is called a unit element in $\widetilde{G}$, if $\gamma^4 = 1$.

There are only four unit elements, 1, i, -1, $-i$. Between any two of them, say $i, -1$, they are divisors of each other. Conventionally, the set of all unit elements is denoted by $\mathbb{U}$. This concept can be extended as follows.

Definition 5.3. Two integers $\alpha, \beta \in \widetilde{G}$ are called associates if $\alpha = \beta\gamma$ and γ is a unit.

For example, there is an associated relationship among $m + in$, $-n + im$, $-m - in$, and $n - im$. Note that in general, the conjugate relation between $m + in$ and $m - in$ is not an associated one.

Definition 5.4. A non-zero, non-unit Gaussian integer $\alpha \in \widetilde{G}$ is called *irreducible*, if the only divisors of α are the unit elements and associates of α.

Oppositely, an integer $\alpha \in \widetilde{G}$ is called reducible in $\widetilde{G}$, if there are other divisors of α in addition to the unit elements and associates of α.

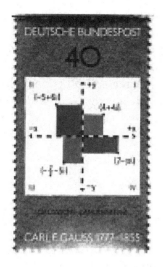 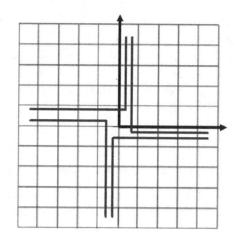

Fig. 5.4 Gaussian quadrant on a postage stamp with details

5.4.3 Unique factorization theorem in $\widetilde{G}$

It is known that there is a unique factorization theorem for any natural number $n > 1$, that is,

$$n = p_1^{r_1(n)} p_2^{r_2(n)} \cdots p_k^{r_k(n)}, \qquad (5.7)$$

where p_i represents prime, and $p_i \neq p_j$ for arbitrary $i \neq j$, and also, $r_i(n) \geq 1$ [2]. Now we shall ask, is there a similar theorem for Gaussian integers? In fact, any non-unit Gaussian integer α in $\widetilde{G}$ can be factorized into irreducible in $\widetilde{G}$ as follows.

$$\alpha = u\, P_1^{k_1} P_2^{k_2} \cdots P_i^{k_i} \cdots P_t^{k_t}, \qquad (5.8)$$

where $P_1, P_2, ..., P_i, ..., P_t$ are distinct irreducible integers which are not associated with each other in $\widetilde{G}$, k_i is positive integer in $\mathbb{N}$. Even though the uncertainty of the order of irreducible integers is ignored in Eq. (5.8), there is still fuzziness caused by associates. In order to eliminate the fuzziness, the quadruple property of any irreducible is removed by the restriction that all the irreducible Gaussian integers are represented in only one quadrant from the four, thus the uniqueness of factorization is ensured[Nar89]. In

---

[2] Note that here $r_i(n)$ is not the coordination number

other words, each irreducible integer must be expressed in a special quadrant. And the unit element u in Eq. (5.8) is determined afterwards. Of course, u is not necessarily restricted in that quadrant since u is not an irreducible integer. If the factorized expansion of a Gaussian integer α only includes one first power of an irreducible as a factor and others are the zero-th powers in Eq. (5.8), then α must be irreducible in $\widetilde{G}$. In order to learn the expansion of $\alpha = m + in$, one has to judge the reducibility of α at the beginning.

5.4.4 Criteria for reducibility

First of all, let us consider a natural prime $p \in \mathbb{N}$, to see the reducibility of p in $\widetilde{G}$. If the prime p is reducible in $\widetilde{G}$, p must take the form as $\alpha\overline{\alpha} = (m+in)(m-in) = m^2 + n^2$.

Definition 5.5. The product of a Gaussian integer α and its conjugate $\overline{\alpha}$ is a sum of two square, called the norm of α, denoted as $\mathcal{N}(\alpha) = \alpha\overline{\alpha}$.

If n is a composite number in $\mathbb{N}$, n must also be reducible in $\widetilde{G}$. The natural numbers can be classified according to composite and prime. It is well known that all the even natural numbers except 2 are composite in $\mathbb{N}$, they must be reducible in $\widetilde{G}$, while by test it is seen that $2 = (1+i)(1-i)$. Thus all the even numbers are reducible in $\widetilde{G}$. Now only the reducibility of odd prime n in $\mathbb{N}$ is explored. Notice that the odd primes can be classified into two types:

$$\begin{cases} 4k+1 \text{ type such as } 5, 13, 17, 29, 37 \\ 4k+3 \text{ type such as } 3, 7, 11, 19, 23 \end{cases}.$$

Now let us explain the difference between the two kinds of odd primes based on congruence analysis, which provide a convenient way to describe divisibility properties. Evidently, $\{m^2 + n^2\}(\mathrm{mod}\ 4)$ only takes 0,1,2 as residues. From multiplicative analysis, we can see that the set of norm satisfies the closeness regarding multiplication shown in Table 5.2.
It is concluded that:

(1) an odd prime of the 4k+3 kind is impossible as a norm, or an odd prime of the 4k+3 kind must be irreducible in $\widetilde{G}$. Hence,

$$|\alpha| = p \equiv 3(\mathrm{mod}\ 4) \Rightarrow \alpha \text{ and its associates are irreducible.}$$

(2) if an odd prime is a norm or a reducible one in $\widetilde{G}$, it must belong to $4k+1$ kind. In other words,

$$\mathcal{N}(\alpha) = p \equiv 1(\mathrm{mod}\ 4) \Rightarrow \alpha \text{ and its associates are irreducible.}$$

This can be proven by contradiction. If $\mathcal{N}(\alpha) = p \equiv 1 (\text{mod } 4)$ and $\alpha = \beta\gamma$ with $|\beta|, |\gamma| > 1$, then $\mathcal{N}(\beta), \mathcal{N}(\gamma) > 1$, and $\mathcal{N}(\alpha) = \mathcal{N}(\beta)\mathcal{N}(\gamma)$ is impossible as a prime.

(3) Note that the congruence analysis also implies that 2 is only even prime being a norm. It is given by

$$\mathcal{N}(\alpha) = p = 2 \Rightarrow \alpha \text{ and its associates are irreducible}$$

Combining the three, an important theorem is given by

Theorem 5.2.

$$\begin{cases} \text{Either } \mathcal{N}(\alpha) = p = 2 \\ \text{or } \mathcal{N}(\alpha) = p \equiv 1 (\text{mod } 4) \\ \text{or } |\alpha| = p \equiv 3 (\text{mod } 4) \end{cases} \Rightarrow \begin{cases} \alpha \text{ and its associates} \\ \text{are irreducible} \end{cases} \quad (5.9)$$

In other words, a Gaussian integer α is irreducible, provided that either the norm of α is 2, or the norm of α is the natural prime of type of $4k+1$, or α is the natural prime of type of $4k+3$ and its associates of the prime.

In fact, the inverse theorem of Eq. (5.9) is also held, thus Eq. (5.9) can be generalized as

Theorem 5.3.

$$\begin{cases} \text{Either } \mathcal{N}(\alpha) = p = 2 \\ \text{or } \mathcal{N}(\alpha) = p \equiv 1 (\text{mod } 4) \\ \text{or } |\alpha| = p \equiv 3 (\text{mod } 4) \end{cases} \Leftrightarrow \begin{cases} \alpha \text{ and its associates} \\ \text{are irreducible} \end{cases} \quad (5.10)$$

This generalized theorem shows that any one of the three cases makes α irreducible, and conversely, once α is irreducible, then there must exist one of the three cases. This theorem becomes a powerful criterion to judge the reducibility of $\alpha = m + in$ in $\tilde{G}$. In this book we just introduce the content of the theorem and its applications, and remove the complete proof since it is easy to find in standard textbooks[Nar89; Pan2001].

Table 5.2 $(m^2 + n^2)(\text{mod } 4)$

$\otimes$	0	1	2
0	0	0	0
1	0	1	2
2	0	2	0

5.4.5 Procedure for factorization into irreducibles

The factorization procedure for an arbitrary Gaussian integer $\alpha = m + in$ can be decomposed into several steps.

(1) Assuming that $\alpha = m + in$ can be factorized as

$$\alpha = u \cdot (1+i)^t \cdot P_1^{K_1} \cdots P_j^{K_j} \cdots P_s^{K_s}. \tag{5.11}$$

Different from Eq. (5.8), the terms related to $1+i$ and its associates have been listed individually, $\{P_j^{K_j}\}$ represents the irreducible factors except $1+i$ and its associates. Here, the representation of all the factors is set in the first quadrant. The choice of u is such that the final α is also standing at the first quadrant.

(2) From Eq. (5.8), a special expansion of the norm of α is obtained as

$$\mathcal{N}(\alpha) \equiv \alpha\overline{\alpha} = 2^t \cdot [P_1\overline{P_1}]^{K_1} \cdots [P_j\overline{P_j}]^{K_j} \cdots [P_s\overline{P_s}]^{K_s} \tag{5.12}$$

This must be an expansion of a natural number. According to the theorem Eq. (5.10), there are two cases for the norm of each irreducible P_j:

(a) $P_j\overline{P_j}$ is a natural prime number of type of $4k+1$, denoted as q_+

(b) $P_j\overline{P_j}$ is a natural composite number of type of $4k+1$, which must be the square of a natural prime of type of $4k+3$, denoted as q_-^2

Since the factorization theorem in $\mathbb{N}$ is well known, it follows that

$$\mathcal{N}(\alpha) = 2^t \cdot \prod_\beta q_{+\beta}^{\kappa_\beta} \cdot \prod_\gamma q_{-\gamma}^{2\kappa_\gamma}. \tag{5.13}$$

(3) Go back from the norm to α itself, it is given by

$$\alpha = u \cdot (1+i)^t \cdot \prod_\beta (m_\beta + in_\beta)^{\kappa_\beta} \cdot \prod_\gamma q_{-\gamma}^{\kappa_\gamma}. \tag{5.14}$$

where $m_\beta + in_\beta$ is in the first quadrant. The unique factorization is completed.

Now let us look at an example for factorization of $\alpha = 972 + 180i$.

(1) Abstracting the common factors between 972 and 180, we have

$$\alpha = 2^2 \cdot 3^2 \cdot (27 + 5i).$$

(2) Taking the norm, then

$$\mathcal{N}(\alpha) = (2^4) \cdot (3^4) \cdot (27+5i)(27-5i) = (2^4) \cdot (3^4) \cdot (754) = 2^5 \cdot 3^4 \cdot 13 \cdot 29.$$

(3) Considering that 3^2 is of the type of $4k+1$, and 3 is a prime of the type of $4k+3$; 13 and 29 both are prime numbers of the type of $4k+1$,

then

$$\begin{cases} 2^4 & \Rightarrow 2^2 \\ 2 & \Rightarrow (1+i) \\ 3^4 & \Rightarrow 3^2 \\ 13 & \Rightarrow (3+2i) \\ 29 & \Rightarrow (5+2i) \end{cases}$$

Therefore,

$$\alpha = u \cdot 2^2 \cdot (1+i) \cdot 3^2 \cdot (3+2i) \cdot (5+2i) = u(-5+27i).$$

Notice that $-5 + 27i$ is not in the first quadrant. u is chosen such that $\alpha = u(-5 + 27i)$ is located in the first quadrant. In other words, $u = -i$. Finally the result of factorization of α is

$$\alpha = (-i) \cdot (2^2) \cdot (1+i) \cdot (3^2) \cdot (3+2i) \cdot (5+2i).$$

5.4.6 Sum rule of Möbius functions and Möbius inverse formula

The set $\widetilde{G}$ of all non-zero Gaussian integers forms a multiplicative semi-group. Similar to the Möbius function on positive integers, the Möbius function on $\widetilde{G}$ can be defined as

Definition 5.6.

$$\mu(\alpha) = \begin{cases} 1, & \text{if } \alpha \in U \\ (-1)^s, & \text{if } \alpha \text{ is a product of s distinct irreducibles} \\ 0, & \text{otherwise} \end{cases} \quad (5.15)$$

Note that the 'distinct irreducible' in this definition means distinct and distinct un-associated irreducible. From the above, the four associated Gaussian integers correspond to the same value of the Möbius functions. By using the procedure in the last section, one may list all the irreducibles in $\widetilde{G}$, as well as the factorizations of all the reducible in $\widetilde{G}$. Then according to Eq. (5.15), the Möbius functions on $\widetilde{G}$ are given as shown in Table 5.3. A large database of the Möbius functions on $\widetilde{G}$ can be found in 'Mathematica'. Therefore, the sum rule of Möbius functions $\mu(\alpha)$ becomes

$$\sum_{\alpha|z} \mu(\alpha) = \begin{cases} 0, & \text{if } z \notin U \\ 4, & \text{if } z \in U \end{cases} \quad (5.16)$$

Table 5.3 Möbius functions $\mu(\alpha) = \mu(m+in)$ on $\widetilde{G}$

$m \backslash n$	0	1	2	3	4	5	6	7	8
1	1	−1	−1	1	−1	1	−1	0	1
2	0	−1	0	−1	0	−1	0	−1	0
3	−1	1	−1	1	0	1	1	1	0
4	0	−1	0	0	0	−1	0	1	0
5	1	1	−1	1	−1	−1	−1	1	−1
6	0	−1	0	1	0	−1	0	1	0
7	−1	0	−1	1	1	1	1	1	−1
8	0	1	0	−1	0	−1	0	−1	0

Based on the sum rule Eq. (5.16), the Möbius inverse formula on $\widetilde{G}$ can be expressed as

$$E(x) = \frac{1}{2} \sum_{(m,n)\neq(0,0)} \Phi(\sqrt{m^2+n^2}\, x) = \frac{1}{2} \sum_{\alpha \neq 0} \Phi(|\alpha|\, x)$$

$$\Longrightarrow \Phi(x) = \frac{1}{8} \sum_{\alpha \neq 0} \mu(\alpha) E(|\alpha|\, x). \quad (5.17)$$

Proof.

$$\frac{1}{8} \sum_{\alpha \neq 0} \mu(\alpha) E(|\alpha|\, x) = \frac{1}{8} \sum_{\alpha \neq 0} \mu(\alpha) [\frac{1}{2} \sum_{\beta \neq 0} \Phi(\alpha\beta\, x)]$$
$$= \frac{1}{16} \sum_{\gamma} [\sum_{\alpha | \gamma} \mu(\alpha)] \Phi(|\gamma|x) = \frac{1}{4} \sum_{\gamma \in U} \Phi(|\gamma|\, x) = \Phi(x). \quad \square$$

In the above deduction, one has to take care of the range of (m, n) and α.

5.4.7 Coordination numbers in 2D square lattice

Let us set atoms in each lattice point of a 2D square lattice, and choose the lattice constant as the unit of length, then we denote the distances between atoms as $b(1), b(2), b(3), ...$, in other words, $b(1)$ is the first neighboring distance (the nearest neighbor distance), $b(2)$ the second neighboring distance, $b(3)$ the third neighboring distance,... . Correspondingly, we have

Table 5.4 Symmetry analysis of N(x+iy)

(x,y)	$0 = x \neq y$	$x \neq y = 0$	$x = y \neq 0$	$0 \neq x \neq y \neq 0$
coordination number	4	4	4	8

the n-th coordination number $r(n)$, which is the number of atoms with the distance $b(n)$ from the reference atom at origin. In fact, the coordination number is just the number of the solutions of equation for non-zero norm $\mathcal{N}(n) = [b(n)]^2$

$$x^2 + y^2 = \mathcal{N} \qquad (5.18)$$

in which x, y are integers. Obviously, both $b(n)$ and $r(n)$ are arithmetic functions. Conventionally, the coordination numbers can be evaluated by symmetry analysis, and the outline of the method is shown in Table 5.4 with the results in Table 5.5.

From the table, we see that the value of each $[b(n)]^2$ is a sum of two squares, this is why in Table 5.3 we did not see a case with $[b(n)]^2 = 3, 6, 7, 11, 12, ...$, since these values are not a norm, or not a sum of two squares. In mathematics, the cases with $[b(n)]^2 = 3, 6, 7, 11, 12, ...$ are also involved regarding to the vanished coordination numbers, that is, $r(n) = 0$.[3] The general formula for $\mathbf{N}(n)$ is

$$\mathbf{N}(n) = 4P(n) = 4\sum_{d|n} h(d) \qquad (5.19)$$

where $h(n)$ is defined as[4]

$$h(n) = \begin{cases} 1, & \text{if } n = 4k+1 \\ -1, & \text{if } n = 4k+3 \\ 0, & \text{if } n = 2k \end{cases} \qquad (5.20)$$

As shown in Table 5.5 and Eq. (5.20), let $\mathbb{N}(n) = [b(n)]^2$ run over all the natural numbers, whether $\mathcal{N}(n)$ is a norm or not. Thus we can establish Table 5.5 as follows.

---
[3]The coordination number is similar to the degeneracy for a physical system. It is well known that the symmetry analysis can only solve the symmetric degeneracy in physics, but not the incident degeneracy. But the elementary theory of numbers can solve both in a unified way.

[4]Note that

$$h(n) = \sum_{d|n} \mu(d)\mathcal{N}(\frac{n}{d})$$

Table 5.5 Distance function and coordination number

$b^2(n)$	1	2	3	4	5	6	7	8	9	10
$h(n)$	1	0	-1	0	1	0	-1	0	1	0
$r(n)$	4	4	0	4	8	0	0	4	4	8
$b^2(n)$	11	12	13	14	15	16	17	18	19	20
$h(n)$	-1	0	1	0	-1	0	1	0	-1	0
$r(n)$	0	0	8	0	0	4	8	4	0	8
$b^2(n)$	21	22	23	24	25	26	27	28	29	30
$h(n)$	1	0	-1	0	1	0	-1	0	1	0
$r(n)$	0	0	0	0	12	8	0	0	8	0

Now let us explain the source of Eqs. (5.19) and (5.20). Equation (5.12) shows that the unique factorization of a norm can be expressed as

$$\mathbb{N}(\alpha) = 2^t \cdot \prod_\beta q_{+\beta}^{\kappa_\beta} \cdot \prod_\gamma q_{-\gamma}^{2\kappa_\gamma}$$

where $q_{+\beta}$ represents a factor of $4k+1$ type, and $q_{-\gamma}$ represents a factor of $4k-1$ type. For our purpose, N should run over all the positive integers as n, hence, Eq. (5.12) will be changed into

$$n = 2^t \cdot \prod_\beta q_{+\beta}^{\kappa_\beta} \cdot \prod_\gamma q_{-\gamma}^{\kappa_\gamma} \tag{5.21}$$

In other words, the expansion of a norm $\mathbb{N}(\alpha)$ includes all the $q_{-\gamma}$ to the power of even number, while the expansion of an arbitrary positive integer n includes all the $q_{-\gamma}$ to the power of either odd or even. Since the factor 2^t is not supposed to contribute the different way to express n as a sum of two squares such that $\mathbb{N}(n_2) = \mathbb{N}(2^t) = 1$, the $\mathbb{N}(n)$ can be simplified as a product of two independent parts for it is multiplicative and $(n_+, n_-) = 1$.

$$\begin{cases} \mathbb{N}(n) = \mathbb{N}(n_+)\mathbb{N}(n_-) = \mathbb{N}(\prod_\beta q_{+\beta}^{\kappa_\beta}) \cdot \mathbb{N}(\prod_\gamma q_{-\gamma}^{2\kappa_\gamma}) \\ n_+ = \prod_\beta q_{+\beta}^{\kappa_\beta} \\ n_- = \prod_\gamma q_{-\gamma}^{2\kappa_\gamma} \end{cases} \tag{5.22}$$

Note that

$$\mathcal{N}(q_\gamma^\kappa) = \begin{cases} 1 & 2|\kappa \\ 0 & 2 \nmid \kappa \end{cases} \tag{5.23}$$

it follows that
$$\mathcal{N}(n_-) = \begin{cases} 1 & n_- \text{ is a square} \\ 0 & n_- \text{ is not a square} \end{cases} \quad (5.24)$$

Therefore,
$$\mathcal{N}(n) = \begin{cases} \mathbb{N}(n_+), & \text{if } n_- \text{ is a square,} \\ 0, & \text{if } n_- \text{ is not a square.} \end{cases} \quad (5.25)$$

In Chapter 1, $\tau_1(n)$ and $\tau_3(n)$ have been defined as the number of factors of n as $4k+1$ and $4k+3$ respectively, and

$$\mathfrak{F}(n) = \tau_1(n) - \tau_3(n) = \sum_{\substack{d|n \\ d \equiv 1 (\bmod\ 4)}} 1 - \sum_{\substack{d|n \\ d \equiv 3 (\bmod\ 4)}} 1 \quad (5.26)$$

Hence
$$\mathfrak{F}(n) = \mathfrak{F}(n_2)\mathfrak{F}(n_+)\mathfrak{F}(n_-) \quad (5.27)$$

Consider that
$$\mathfrak{F}(n_2) = \tau_1(n_2) - \tau_3(n_2) = 1 - 0 = 1 \quad (5.28)$$

thus
$$\mathfrak{F}(n) = \mathfrak{F}(n_+ n_-) = \mathfrak{F}(n_+)\mathfrak{F}(n_-) \quad (5.29)$$

Note that
$$\tau_1(q_-^\kappa) = [\frac{\kappa}{2}] + 1$$
$$\tau_3(q_-^\kappa) = [\frac{\kappa+1}{2}]$$

where $[x]$ is the integer part of x. Hence,

$$\mathfrak{F}(q_-^\kappa) = \tau_1(q_-^\kappa) - \tau_3(q_-^\kappa) = \begin{cases} 1 & \text{if } \kappa = 2s \\ 0 & \text{if } \kappa = 2s-1 \end{cases}$$

and
$$\mathfrak{F}(n_-) = \begin{cases} 1 & \text{if } n_- \text{ is a square} \\ 0 & \text{if } n_- \text{ is not a square} \end{cases} \quad (5.30)$$

From Eqs. (5.29) and (5.30), it follows that

$$\mathfrak{F}(n) = \begin{cases} \mathfrak{F}(n_+) & \text{if } n_- \text{ is a square} \\ 0 & \text{if } n_- \text{ is not a square} \end{cases} \quad (5.31)$$

Considering the definition of $h(n)$, it follows that

$$\mathfrak{F}(n) = \tau_1(n) - \tau_3(n) = \sum_{d|n} h(d) \qquad (5.32)$$

Notice that

$$\tau_1(q_+^\beta) = \beta + 1$$
$$\tau_3(q_+^\beta) = 0$$

then

$$\mathfrak{F}(q_+^\beta = \beta + 1.$$

Hence,

$$\mathfrak{F}(\prod q_+^\beta) = \prod_j (\beta_j + 1) \qquad (5.33)$$

Considering the four quadrants, we have

$$\mathcal{N}(n) = 4\prod_j (\beta_j + 1) = 4\tau(n_1) = 4[\tau_1(n) - \tau_3(n)] = 4\sum_{d|n} h(d) \qquad (5.34)$$

This is the source of Eqs. (5.19) and (5.20).

5.4.8 Application to the 2D arithmetic Fourier transform

The discrete Fourier transform (DFT) is an essential part of numerous physical and technical problems such as band structure, signal processing, pattern recovery, etc. The popularity of the fast Fourier transform (FFT) is primarily due to its reduction of the number of multiplicative operations from $O(N^2)$ to $O(N \log N)$ for computing $2N + 1$ Fourier coefficients. The arithmetic Fourier transform(AFT)[12] is based on the Möobius inversion formula of classical number theory[1], in which the number of multiplications is further reduced significantly down to $O(N)$ since the Möbius functions only take the value in $\{-1, 0, 1\}$. At the same time, the remaining addition operations can be performed much more quickly by a parallel processor.

A new 2D AFT algorithm has been proposed by Tufts et al.[Tuf89]. Also, arrow-column algorithm has been developed by Kelly and co-workers[Kel93]. All of those methods require that the row, column, and the global means of the function must be removed before transforming. In the present work, a new 2D AFT technique based on the Möbius inversion formula on Gaussian integers is introduced in a concise manner. Suppose that a function $f(x,y)$ is defined on R^2 with a period of 2π for both the x and y, and with the zero "direct component", that is,

$$f(x,y) = \sum_{(m,n)\neq(0,0)} c_{m,n} e^{i(mx+ny)} \tag{5.35}$$

with

$$\int_D f(x,y)\, dx\, dy = \int_0^{2\pi}\int_0^{2\pi} f(x,y)\, dx\, dy = 0 \tag{5.36}$$

where $D = [0, 2\pi] \times [0, 2\pi]$, and

$$c_{m,n} = \frac{1}{4\pi^2}\int_D f(x,y) e^{-i(mx+ny)}\, dx\, dy \tag{5.37}$$

Now let us use the notation $\alpha = m + ni$, and write $c_{m,n} = c_\alpha$. The sum in Eq. (5.34) is equivalent to a sum over α, with α taking all values in the set $\widetilde{G}$ of non-zero Gaussian integers. As discussed before, there is a unique factorization property in Gaussian integers and a Möbius inversion formula, which will be used below.

Writing $\beta = a + bi \in \widetilde{G}$, we have

$$S_\alpha \equiv \sum_{\beta \in \widetilde{G}} c_{\alpha\beta} = \frac{1}{4\pi^2}\int_D [\sum_{\beta \in \widetilde{G}} e^{-i[(am-bn)x+(bm+an)y]}] f(x,y) dx dy$$

$$= \frac{1}{4\pi^2}\int_D [\sum_{\beta \in \widetilde{G}} e^{-ia(mx+ny)} e^{-ib(-nx+my)}] f(x,y) dx dy. \tag{5.38}$$

Note that

$$\sum_{n=-\infty}^{\infty} e^{inx} = 2\pi \sum_{q=-\infty}^{\infty} \delta(x - 2q\pi) \tag{5.39}$$

and considering $\beta = a + bi$, it is given by

$$S_\alpha = \frac{1}{4\pi^2}\int_D [-1 + 4\pi^2 \sum_{p,q=-\infty}^{\infty} \delta(mx+ny-2p\pi)\delta(-nx+my-2q\pi)] f(x,y)dxdy$$

$$= \sum_{p,q=-\infty}^{\infty} \int_D \delta(mx+ny-2p\pi)\delta(-nx+my-2q\pi) f(x,y) dx dy \tag{5.40}$$

Since D is a bounded region, when $|p|$ and $|q|$ are large enough, the corresponding terms vanish, thus only finitely many terms make a contribution to the above summation. We introduce the transformation of variables

$$\begin{cases} u = mx + ny \\ v = -nx + my \end{cases} \quad (5.41)$$

The corresponding Jacobian can be expressed as

$$\Delta = \begin{vmatrix} m & n \\ -n & m \end{vmatrix} = m^2 + n^2 = |\alpha|^2 \neq 0 \quad (5.42)$$

and

$$\begin{cases} x = \dfrac{mu - nv}{\Delta} \\ y = \dfrac{nu + mv}{\Delta} \end{cases} \quad (5.43)$$

Therefore, Eq. (5.40) becomes

$$S_\alpha = \sum_{p,q=-\infty}^{\infty} \int_{D'} \frac{\delta(u - 2p\pi)}{\Delta} \delta(v - 2q\pi) f\left(\frac{mu - nv}{\Delta}, \frac{nu + mv}{\Delta}\right) dudv$$

$$= \frac{1}{\Delta} \sum_{(u,v) \in D'(\alpha)} f\left(\frac{mu - nv}{\Delta}, \frac{nu + mv}{\Delta}\right)$$

$$= \frac{1}{\Delta} \sum_{(x,y) \in D(\alpha)} f(x,y) = \frac{1}{|\alpha^2|} \sum_{(x,y) \in D(\alpha)} f(x,y) \quad (5.44)$$

where $D(\alpha)$ is determined uniquely by $\alpha = m + in$. In other words, for certain (m, n), there are only a finite number of points within the region of $[0, 2\pi) \cdot [0, 2\pi)$. These points can be represented by

$$(2\pi(mp - nq)/[m^2 + n^2], 2\pi(np + mq)/[m^2 + n^2])$$

where p, q are integers. The set of finite points is also called interpolation point set.

For the equation

$$S_\alpha = \sum_{\beta \in G^*} c_{\alpha\beta} \quad (5.45)$$

we might use the Möbius inversion formula as

$$\sum_{z \in U} c_{z\alpha} = \frac{1}{4} \sum_{\beta \in G^*} \mu(\beta) S_{\alpha\beta} \quad (5.46)$$

since the right-hand side of Eq. (5.46) is equal to

$$W_1 \equiv \frac{1}{4}\sum_{\beta \in G^*} \mu(\beta) S_{\alpha\beta} = \frac{1}{4}\sum_{\beta \in G^*} \mu(\beta) \sum_{\gamma} c_{\alpha\beta\gamma}$$

$$\stackrel{\beta'=\beta\gamma}{=} \frac{1}{4}\sum_{\beta' \in G^*} c_{\alpha\beta'} \sum_{\beta|\beta'} \mu(\beta) = 4(\frac{1}{4})\sum_{\beta' \in U} c_{\alpha\beta'}$$

$$= c_\alpha + c_{i\alpha} + c_{-\alpha} + c_{-i\alpha} = \frac{1}{4}\sum_{\beta \in G^*} \frac{\mu(\beta)}{|\alpha\beta|^2} \sum_{(x,y) \in D(\alpha\beta)} f(x,y) \quad (5.47)$$

Now let us consider

$$F(x,y) = f(x+s\pi, y+t\pi)$$

$$= \sum_{m,n=-\infty}^{\infty} c_{mn} e^{i[m(x+s\pi)+n(y+t\pi)]}$$

$$= \sum_{m,n=-\infty}^{\infty} c_{mn} e^{i(ms+nt)\pi} \, e^{i(mx+ny)} \quad (5.48)$$

For the periodicity of $f(x,y)$, as well as $c_{00} = 0$, it is given by

$$\int_0^{2\pi}\int_0^{2\pi} F(x,y) dx dy = 0 \quad (5.49)$$

In addition, let $\alpha = m+in$, then from Eq. (5.47) we have

$$c_\alpha e^{i(ms+nt)\pi} + c_{\alpha i} e^{i(-ns+mt)\pi} + c_{-\alpha} e^{i(-ms-nt)\pi} + c_{-\alpha i} e^{i(ns-mt)\pi}$$

$$= \frac{1}{4}\sum_{\beta \in G^*} \frac{\mu(\beta)}{|\alpha\beta|^2} \sum_{(x,y) \in D(\alpha\beta)} f(x+s\pi, y+t\pi) \quad (5.50)$$

(i) Let $s = \frac{n}{\Delta}, t = \frac{-m}{\Delta}$, then

$$W_2 \equiv c_\alpha e^{i0\pi} + c_{i\alpha} e^{-i\pi} + c_{-\alpha} e^{i0\pi} + c_{-i\alpha} e^{i\pi} = c_\alpha - c_{i\alpha} + c_{-\alpha} - c_{-i\alpha}$$

$$= \frac{1}{4}\sum_{\beta' \in G^*} \frac{\mu(\beta)}{|\alpha\beta|^2} \sum_{(x,y) \in D(\alpha\beta)} f(x + \frac{n}{m^2+n^2}\pi, y + \frac{-m}{m^2+n^2}\pi). \quad (5.51)$$

(ii) Let $s = \frac{m+n}{2\Delta}, t = \frac{-m+n}{2\Delta}$, then

$$W_3 \equiv i[c_\alpha + c_{i\alpha} - c_{-\alpha} - c_{-i\alpha}] = \frac{1}{4}\sum_{\beta' \in G^*} \frac{\mu(\beta)}{|\alpha\beta|^2} \times$$

$$\times \sum_{(x,y) \in D(\alpha\beta)} f(x + \frac{m+n}{2(m^2+n^2)}\pi, y + \frac{-m+n}{2(m^2+n^2)}\pi) \quad (5.52)$$

(iii) Let $s = \frac{m-n}{2\Delta}, t = \frac{m+n}{2\Delta}$, then

$$W_4 \equiv i[c_\alpha - c_{i\alpha} - c_{-\alpha} + c_{-i\alpha}] = \frac{1}{4} \sum_{\beta' \in G^*} \frac{\mu(\beta)}{|\alpha\beta|^2} \times$$

$$\times \sum_{(x,y) \in D(\alpha\beta)} f(x + \frac{m-n}{2(m^2+n^2)}\pi, y + \frac{m+n}{2(m^2+n^2)}\pi). \quad (5.53)$$

From the above, we have $c_{i\alpha} - c_{-i\alpha} = \frac{i}{2}(W_2 - W_3)$ and so on. Finally, we have

$$\begin{cases} c_\alpha = \frac{1}{4}[(W_1 + W_2) - i(W_3 + W_4)] \\ c_{-\alpha} = \frac{1}{4}[(W_1 + W_2) + i(W_3 + W_4)] \\ c_{i\alpha} = \frac{1}{4}[(W_1 - W_2) + i(W_3 - W_4)] \\ c_{-i\alpha} = \frac{1}{4}[(W_1 - W_2) - i(W_3 - W_4)] \end{cases} \quad (5.54)$$

In most cases, $f(x,y)$ is real, and all the W_1, W_2, W_3 and W_4 are real, thus $\overline{c_\alpha} = c_{-\alpha}$. From the above we see that the multiplicative operations in the AFT procedure are much less than that in FFT. This makes AFT quite available for a parallel processing design. However, the sampling distribution needs further improvement for practical application.

The above method requires to remove all the direct components (the row average, column average and global average) before carrying on the Fourier transform. In Chapter 4, it was mentioned that the direct component can be removed in AFT by using Bruns method. Now let us consider the Bruns version for 2D AFT.

5.4.9 Bruns version of 2D AFT and VLSI architecture

5.4.9.1 The derivation of Bruns version for 2D AFT

Now, let us recall the Bruns version of 1D AFT in Chapter 4 first. In other words, let us review Eqs. (4.34)–(4.37) as follows.

$$a(0) = \frac{1}{2\pi} \int_0^{2\pi} f(x)dx$$

$$a(n) = \sum_{l=1,3,5,\cdots}^{[N/n]} \mu(l) B(2nl, 0)$$

$$b(n) = \sum_{l=1,3,5,\cdots}^{[N/n]} \mu(l)(-1)^{(l-1)/2} B(2nl, \frac{1}{4nl})$$

where the alternating average $B(2n, \alpha)$ of the periodic function $f(x)$ with phase shift is

$$B(2n, \alpha) = \frac{1}{2n} \sum_{m=1}^{2n} (-1)^m f\left(2\pi \frac{m}{2n} + 2\pi\alpha\right).$$

Equation (4.37) can be also expressed as

$$B(2n, \gamma) = \frac{1}{2n} \sum_{l=1}^{2n-1} (-1)^l f\left(\frac{l}{2n}T + \gamma T\right).$$

Now let us consider the Bruns version for 2D AFT. The corresponding equation can be rewritten as

$$\begin{aligned}A(x,y) = \sum_{m=0}^{M} \sum_{n=0}^{N} [&a_{m,n} \cos(m\omega_1 x) \cos(n\omega_2 y) \\ &+ b_{m,n} \sin(m\omega_1 x) \sin(n\omega_2 y) \\ &+ c_{m,n} \sin(m\omega_1 x) \cos(n\omega_2 y) \\ &+ d_{m,n} \cos(m\omega_1 x) \sin(n\omega_2 y)\end{aligned} \quad (5.55)$$

where $m = 1, 3, 5, ..., M$, $n = 1, 3, 5, ..., N$. Similar to Eqs. (4.35) and (4.36), $a_{m,n}, b_{m,n}, c_{m,n}$ and $d_{m,n}$ can be determined as

$$a_{m,n} = \sum_{p=1,3,...}^{[M/m]} \sum_{q=1,3,5,...}^{[N/n]} \mu(p)\mu(q) B(2mp, 0; 2nq, 0) \quad (5.56)$$

$$b_{m,n} = \sum_{p=1,3,..}^{[M/m]} \sum_{q=1,3,..}^{[N/n]} \mu(p)\mu(q)(-1)^{\frac{p+q-2}{2}} B\left(2mp, \frac{1}{4mp}; 2nq, \frac{1}{4nq}\right) \quad (5.57)$$

$$c_{m,n} = \sum_{p=1,3,5,...}^{[M/m]} \sum_{q=1,3,5,...}^{[N/n]} \mu(p)\mu(q)(-1)^{\frac{(p-1)}{2}} B\left(2mp, \frac{1}{4mp}; 2nq, 0\right) \quad (5.58)$$

$$d_{m,n} = \sum_{p=1,3,5,...}^{[M/m]} \sum_{q=1,3,5,...}^{[N/n]} \mu(p)\mu(q)(-1)^{\frac{(q-1)}{2}} B\left(2mp, 0; 2nq, \frac{1}{4nq}\right) \quad (5.59)$$

in which, similar to (4.37), $B(2m, \alpha_1; 2n, \alpha_2)$ is defined as[Ge97]

$$B(2m, \gamma_1; 2n, \gamma_2) = \sum_{p=0}^{2m-1} \sum_{q=0}^{2n-1} \frac{(-1)^{p+q}}{4mn} A\left(\left(\frac{p}{2m}+\gamma_1\right)T_1; \left(\frac{p}{2n}+\gamma_2\right)T_2\right) \quad (5.60)$$

where $(\gamma_1, \gamma_2) \in [0,1)$ are phase factors, and m,n are positive integers.

According to the definition, the row, column, and global means have already been removed from $B(2m, \gamma_1; 2n, \gamma_2)$ by the factor $(-1)^{p+q}$.

If we write the 2D periodic function as

$$A(x,y) = \sum_{m=-M}^{M} \sum_{n=-N}^{N} \rho_{m,n} exp[i(m\omega_1 x + n\omega_2 y)] \qquad (5.61)$$

where $\omega_1 = 2\pi/T_1, \omega_2 = 2\pi/T_2$ and

$$\rho_{m,n} = \frac{1}{T_1 T_2} \int_0^{T_1} \int_0^{T_2} A(x,y) exp[-i(m\omega_1 x + n\omega_2 y)] dx dy \qquad (5.62)$$

then there are some relations between $a_{m,n}, b_{m,n}, c_{m,n}, d_{m,n}$ and $\rho_{m,n}$ as follows.

$$\begin{pmatrix} \rho_{m,n} \\ \rho_{-m,-n} \\ \rho_{-m,n} \\ \rho_{m,-n} \end{pmatrix} = \frac{1}{4} \begin{pmatrix} 1 & -1 & -i & -i \\ 1 & -1 & i & i \\ 1 & 1 & i & -i \\ 1 & 1 & -i & i \end{pmatrix} \begin{pmatrix} a_{m,n} \\ b_{m,n} \\ c_{m,n} \\ d_{m,n} \end{pmatrix} \qquad (5.63)$$

in particular, when $m = 0$ and $n \neq 0$, or when $n = 0$ and $m \neq 0$

$$\begin{pmatrix} \rho_{0,n} \\ \rho_{0,-n} \end{pmatrix} = \frac{1}{2} \begin{pmatrix} 1 & -i \\ 1 & i \end{pmatrix} \begin{pmatrix} a_{0,n} \\ d_{0,n} \end{pmatrix} \qquad (5.64)$$

and

$$\begin{pmatrix} \rho_{m,0} \\ \rho_{-m,0} \end{pmatrix} = \frac{1}{2} \begin{pmatrix} 1 & -i \\ 1 & i \end{pmatrix} \begin{pmatrix} a_{m,0} \\ c_{m,0} \end{pmatrix} \qquad (5.65)$$

Note that in the Bruns version of 2D AFT, the sum runs over only the odd numbers, therefore, there is no information related to $\rho_{0,k}$ and $\rho_{k,0}$, which correspond to row and column means respectively and require individual treatment[Ge97].

Let $B_1(2m, \gamma_1)$ be the 2mth average of A(x,y) with respect to x

$$B_1(2m, \gamma_1) = \frac{1}{4mL_2} \sum_{p=0}^{2m-1} \sum_{q=0}^{2L_2-1} (-1)^p A\left(\frac{p}{2m}T_1 + \gamma_1 T_1, \frac{q}{2L_2}T_2\right) \qquad (5.66)$$

where $L_2 > N/2$.

In fact, $B_1(2m, \gamma_1)$ is independent of L_2, and therefore, we can set $L_2 = N/2 + 1$. It can be proven that $B_1(2m, \gamma_1) = 0$ if $m > M$, and

$$B_1(2m, \gamma_1) = \sum_{p=1,3,5,\ldots}^{[M/m]} g_{mp,0}(\gamma_1, 0) \qquad (5.67)$$

Then, based on the 1D Möbius inversion formula, for any $m = 1, 2, ..., M$, it is given by

$$g_{m,0}(\gamma_1, 0) = \sum_{p=1,3,5,...}^{[M/n]} \mu(p) B_1(2mp, \gamma_1) \text{ for n=1,2,...,N} \quad (5.68)$$

By a parallel way, one gets

$$g_{0,n}(0, \gamma_2) = \sum_{q=1,3,5,...}^{[M/n]} \mu(q) B_1(2nq, \gamma_2) \text{ for n=1,2,...,N} \quad (5.69)$$

where $B_2(2n, \gamma_2)$ is defined in Definition 3.

Let $B_2(2n, \gamma_2)$ be the $2n$-th average of $A(x, y)$ with respect to y

$$B_2(2n, \gamma_2) = \frac{1}{4L_1 n} \sum_{p=0}^{2L_1-1} \sum_{q=0}^{2n-1} (-1)^q A\left(\frac{p}{2L_1} T_1, \frac{q}{2n} T_2 + \gamma_2 T_2\right) \quad (5.70)$$

where $L_2 > M/2$. From Eqs. (5.68) and (5.69), by choosing suitable γ_1 and γ_2, one gets $a_{0,n}, a_{n,0}, c_{m,0},$ and $d_{0,n}$.

For $(1 \leq m \leq M, 1 \leq n \leq N$, it is given by

$$a_{m,0} = \sum_{p=1,3,5,\cdots}^{[M/m]} \mu(p) B_1(2mp, 0), \quad (5.71)$$

$$c_{m,0} = \sum_{p=1,3,5,\cdots}^{[M/m]} \mu(p)(-1)^{(p-1)/2} B_1\left(2mp, \frac{1}{4mp}\right), \quad (5.72)$$

$$a_{0,n} = \sum_{q=1,3,5,\cdots}^{[N/n]} \mu(q) B_1(2nq, 0), \quad (5.73)$$

$$d_{0,n} = \sum_{q=1,3,5,\cdots}^{[N/n]} \mu(q)(-1)^{(q-1)/2} B_2\left(2nq, \frac{1}{4nq}\right) \quad (5.74)$$

Finally, $a_{m,n}, b_{m,n}, c_{m,n},$ and $d_{m,n}$ for $(0 \leq m \leq M; 0 \leq n \leq N)$ are all obtainable. Then, $\rho_{m,n}$ for $(-M \leq m \leq M, -N \leq n \leq N)$ can be determined.

Note that in this procedure no complex multiplication is involved even for a complex-valued function. Only complex additions and a small number of scaling by real integers are needed, which will lead to a powerful fast Fourier transform algorithm.

Table 5.6 Relative root-mean-square error (RMSE) between input Gaussian function and the reconstructed wave form with a different number of sampling points

Number of samples	256× 256	512× 512	1024× 1024
RMSE	8.1×10^{-4}	4.7×10^{-4}	1.7×10^{-4}

5.4.9.2 Example for 2D AFT computation

A VLSI architecture is presented in Fig. 5.5. First, the 2-D signal is transferred to a uniformly spaced 2-D data array, and the alternative averages are calculated by definitions in Eqs. (5.60), (5.68), and (5.70). Then, $a_{m,n}, b_{m,n}, c_{m,n}$, and $d_{m,n}$ are obtained according to Eqs. (5.56)–(5.59) and Eqs. (5.71)–(5.74). Finally, $\rho_{m,n}$ are derived from $a_{m,n}, b_{m,n}, c_{m,n}$, and $d_{m,n}$ by Eqs. (5.63)–(5.65).

As an example of the present algorithm, we calculate the transform of the real-valued periodic Gaussian function

$$G(x,y) = \frac{1}{2\pi\sigma^2} exp\{-\frac{(x-10)^2 + (y-10)^2}{2\pi\sigma^2}\} \quad (5.75)$$

where $\sigma = 1, T_1 = T_2 = 20$. The algorithm is performed by a computer program. Zero-order interpolation with different number of sampling points is used. Letting M=N=20, the Fourier transform of Gaussian function can be quickly obtained, and the waveform can be successfully reconstructed by inverse Fourier transform. The relative root-mean-square error (RMSE) between the reconstructed wave form and the input function $A(x,y)$ with a different number of sample points is shown in Table 5.6. The error is acceptable for some applications of signal processing. Higher accuracy can be expected by use of the first-order interpolation[Ree90].

In Table 5.7, the algorithm proposed in this correspondence is compared with Tufts' method[Tuf89] and Kelley's row-column method [Kel93].The present algorithm needs more additions than that in [Tuf89] and [Kel93]. However, fewer multiplications are taken in this new method. Kelley's method requires fewer additions and more multiplications. Because the number of sample points is in direct proportion to the number of additions, the present algorithm needs more sample points. In terms of the calculation of $\rho_{k,0}$ and $\rho_{0,k}$, this algorithm has advantages over the methods in [Tuf89] and [Kel93]. The row, column, and global means in Tufts' method must be removed before transform. This is time consuming and difficult for practical applications. The mean-removal step is self-included in the present algorithm.

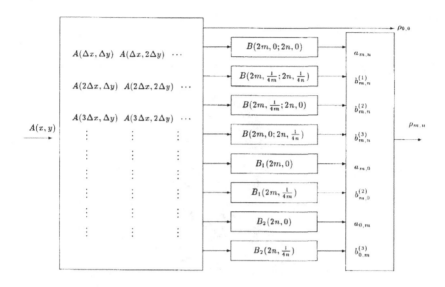

Fig. 5.5 VLSI architecture for the present 2D AFT

Table 5.7 Comparison of different methods for calculating a K-point 2D AFT

	Row-column method in [Kel93]	Method in [Tuf89]	Present method
Additions	$O(K^{3/2})$	$\frac{9}{64}K^2$	$\frac{1}{4}K^2$
Multiplications	$3K$	$\frac{9}{4}K$	K
Complexity	Medium	High	Low

5.5 2D Hexagonal Lattice and Eisenstein Integers

5.5.1 *Definition of Eisenstein integers*

Definition 5.7. Define the Eisenstein integer as

$$\mathbf{E} = a + b\omega | a, b \in \mathbf{Z} \tag{5.76}$$

where a, b are integers, and

$$\omega = \frac{1 + i\sqrt{3}}{2} = e^{i\pi/3}. \tag{5.77}$$

Note that conventionally the Gaussian integers are denoted as $Z[i]$, and the Eisenstein integers as $Z[\omega]$[Sch90; Nar89]. From

$$\omega^2 = e^{2i\pi/3} = e^{i\pi/3} - 1 = \omega - 1$$

it follows that

$$(a+b\omega)(c+d\omega) = ac + (ad+bc)\omega + bd\omega^2 = (ac-bd) + (ad+bc+bd)\omega.$$

In other words, regarding multiplicative operation $Z[\omega]$ is a closed set

Fig. 5.6 Gotthold Eisenstein (1823–1852)

5.5.2 Norm and associates of an Eisenstein integer

For any $\alpha = a + b\omega \in Z[\omega]$, we define the norm of α as

$$N(\alpha) = \alpha \cdot \overline{\alpha} = a^2 + b^2 + ab. \tag{5.78}$$

For example, for a=b=1, we have $\alpha = 1 + [1+\sqrt{3}i]/2$, then $N(\alpha) = 3$. In fact,

$$\alpha \cdot \overline{\alpha} = \{[\frac{3}{2} + \sqrt{3}i]/2\} \cdot \{[\frac{3}{2} - \sqrt{3}i]/2\} = \frac{9}{4} + \frac{3}{4} = 3.$$

Similar to Gaussian integers, a product of two Eisenstein integers is still an Eisenstein integer. The norm has similar property.

The concepts of associates, unit elements, and factorization in $Z[i]$ have their correspondence in $Z[\omega]$, now the unit elements become

$$\mathbf{U} = \pm 1, \pm \omega, \pm \omega^2 \tag{5.79}$$

or $\mathbf{U} = 1,\ e^{i\pi/3},\ e^{2i\pi/3},\ -1,\ e^{4i\pi/3},\ e^{5i\pi/3}$.

Definition 5.8. Two Eisenstein integers have associated relation if their difference is only a factor of unit element. For example, for the set of $\alpha,\ \alpha e^{i\pi/3},\ \alpha e^{2i\pi/3},\ -\alpha,\ \alpha e^{4i\pi/3},\ \alpha e^{5i\pi/3}$ there exists associated relations between any two of them. An Eisenstein integer α is called irreducible in $Z[\omega]$ if the only factors of α are unit elements and its associates. If α has factors different from its associates and unit elements, α is called reducible in $Z[\omega]$.

A non-zero $\alpha \in Z[\omega]$ can be also factorized into a product of irreducibles as
$$\alpha = uP_1^{k_1} P_2^{k_2} ... P_i^{k_i} ... P_t^{k_t} \tag{5.80}$$
where $P_1, P_2, ..., P_i, ..., P_t$ are distinct irreducibles in $\mathbf{Z}[\omega]$, and they are not associated with one another, the positive number k_i represents the power of P_i. As before, the order of these irreducibles is ignored, or the different order causes nothing here. Also, the repeat attributed to associates is eliminated. Thus, it can be shown that the factorization is unique[Pan2001]. In order to judge if an Eisenstein integer $\alpha = m + n\omega$ is reducible or irreducible, the first thing is to judge is that a prime in N is reducible or irreducible in $Z[\omega]$.

5.5.3 Reducibility of an Eisenstein integer

First, let us look which kind of natural numbers is possible to become irreducible in Eisenstein integers. Congruence analysis can accelerate the processing from infinite to finite. From Table 5.8 we see that if the norm $N(\alpha)$ of the Eisenstein integer $\alpha = a + b\omega$ is a prime in N, then the norm must be either $p = 3 = (1 + e^{\frac{\pi}{3}})(1 + e^{-\frac{\pi}{3}})$ or $p \equiv 1 (\bmod\ 3)$, it is impossible to be $p \equiv 2 (\bmod\ 3)$.

The relation between the irreducible in $Z[\omega]$ and the prime in N can be expressed as a theorem as follows.

Theorem 5.4.

$$\begin{cases} N(\alpha) = p = 3 \\ \text{or} \\ N(\alpha) = p \equiv 1 (\bmod\ 3) \\ \text{or} \\ |\alpha| = p \equiv 2 (\bmod\ 3) \end{cases} \Leftrightarrow \alpha \text{ is irreducible in } Z[\omega] \tag{5.81}$$

The above equivalent relation is very important for establishing the Möbius function in $Z[\omega]$. The details can be found in several textbooks[Pan2001; Nar89].

Table 5.8 Congruences analysis N(mod 3) of reducibility

m	n	m^2	n^2	mn	(m^2+n^2+mn)	type	norm is a prime
0	0	0	0	0	0	3k	impossible
1	0	1	0	0	1	3k+1	possible
1	1	1	1	1	0	3k	impossible except p=3
2	0	1	0	0	1	3k+1	possible
2	1	1	1	2	1	3k+1	possible
2	2	1	1	1	0	3k	impossible

5.5.4 Factorization procedure of an arbitrary Eisenstein integer

The last subsection provided a criterion on the reducibility of an arbitrary Eisenstein integer. Now we present a concrete procedure to factorize a reducible Eisenstein integer $\alpha = m + n\omega$. By using Theorem 5.4 the procedure can be expressed as follows.

First, similar to the case of $Z[i]$, we separate $1+\omega$ with its associates from other irreducible in $Z[\omega]$, then

$$\alpha = u \cdot (1+\omega)^t \cdot P_1^{K_1} \cdots P_j^{K_j} \cdots P_s^{K_s} \qquad (5.82)$$

where $\{P_j^{K_j}\}$ represents the set of irreducible factors in $Z[\omega]$ except $1+\omega$, and assuming that all these factors are in the first quadrant. The appearance of the additional u ensures α being at the first quadrant too. Therefore,

$$N(\alpha) \equiv \alpha\overline{\alpha} = 3^t \cdot [P_1\overline{P_1}]^{K_1} \cdots [P_j\overline{P_j}]^{K_j} \cdots [P_s\overline{P_s}]^{K_s} \qquad (5.83)$$

must be a natural number.

Second, in Eq. (5.83), there are two situations to be separated.

(1) $P_j\overline{P_j}$ is a prime in N with type of 3k+1, and denoting $P_j\overline{P_j}$ as q_+

(2) $P_j\overline{P_j}$ is a composite in N with type of 3k+1, and this composite is a square of a prime with type of 3k+2, and denoting $P_j\overline{P_j}$ as q_-^2.

Therefore,

$$N(\alpha) = 2^t \cdot \prod_\beta q_{+\beta}^{\kappa_\beta} \cdot \prod_\gamma q_{+\gamma}^{2\kappa_\gamma} \qquad (5.84)$$

Finally,

$$\alpha = u \cdot (1+\omega)^t \cdot \prod_\beta (m_\beta + n_\beta \omega)^{\kappa_\beta} \cdot \prod_\gamma q_{+\gamma}^{\kappa_\gamma} \qquad (5.85)$$

Table 5.9 Möbius functions on Eisenstein integers

$m \setminus n$	0	1	2	3	4	5	6	7	8
1	1	−1	−1	−1	1	−1	−1	1	−1
2	−1	−1	1	−1	1	1	1	−1	−1
3	0	−1	−1	0	−1	0	0	−1	−1
4	0	1	1	−1	0	−1	1	1	0
5	−1	−1	1	0	−1	−1	1	−1	1
6	0	−1	1	0	1	1	0	−1	1
7	1	1	−1	−1	1	−1	−1	−1	0
8	0	−1	−1	−1	0	1	1	0	0

Notice that all of $(1+\omega)$, $(m_\beta + n_\beta \omega)$ and $q_{+\gamma}$ are in the first quadrant in $Z[\omega]$, which is similar to the first quadrant in $Z[\omega]$, the choice of u in Eq. (5.85) will make α locate at the first sextant. Here we set m=n≠ 1 to avoid repetition with $1+\omega$. From the above, the unique factorization of an arbitrary Eisenstein integer $\alpha = m + n\omega$ can be performed. Then, the construction of Möbius function on $Z[\omega]$ can be done.

5.5.5 Möbius inverse formula on Eisenstein integers

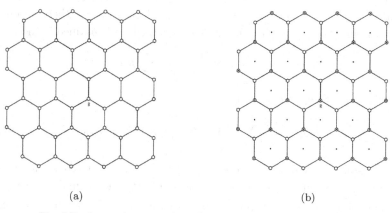

(a) (b)

Fig. 5.7 A monolayer graphite sheet and the Eisenstein lattice

The 2D hexagonal lattice such as a monolayer graphite is essentially a complex lattice as shown in the left part of Fig. 5.7. Rotate it through 180° around a reference point R, then superimpose the new pattern on the origin alone. Obviously, the final result is equivalent to a superposition of two face-centered hexagonal lattices with lattice constants x and $\sqrt{3}x$ respectively(see the right part of Fig. 5.7), and each of them can be represented as a ring of Eisenstein integers, i.e.,

$$\mathcal{E} = \{a + b\omega | m, n \in Z\} \tag{5.86}$$

where m, n are integers in $\mathbb{Z}^5$, and

$$\omega = \frac{1 + i\sqrt{3}}{2} = e^{i\frac{\pi}{3}} \tag{5.87}$$

Since $\omega^2 = \omega - 1$, the product of two Eisenstein integers is itself an Eisenstein integer. The idea of factorization, of units, and of association of elements go through as for the Gaussian integers, except that the units are

$$\mathbb{U} = \{\pm 1, \pm \omega, \pm \omega^2\} \tag{5.88}$$

which represents all the roots of $z^6 = 1$. Any non-zero Eisenstein integer is therefore associated with five other distinct Eisenstein integers.

This is also a multiplicative semigroup with unique factorization considering the associative relations. Therefore, the cohesive energy of the hexagonal lattice is equal to

$$E(x) = \frac{1}{2} \sum_{\alpha \in \mathcal{E}^*} \frac{1}{2} [\Phi(|\alpha| x) + \Phi(\sqrt{3} |\alpha| x)] \tag{5.89}$$

where $\alpha = a + b\omega$ represents an Eisenstein integer, and $\mathcal{E}^*$ is the set of all non-zero Eisenstein integers. Based on the Möbius inverse theorem on the Eisenstein integers, the solution of Eq. (5.89) can be obtained immediately

$$\Phi(|\alpha| x) + \Phi(\sqrt{3} |\alpha| x) = \frac{4}{6^2} \sum_{\alpha \in \mathcal{E}^*} \mu(\alpha) E(|\alpha| x) \tag{5.90}$$

Similar as before, it follows that

$$\Phi(x) = \frac{1}{9} \sum_{n=0}^{\infty} (-1)^n \sum_{\alpha \in \mathcal{E}^*} \mu(\alpha) E(3^{n/2} |\alpha| x). \tag{5.91}$$

This result is easy for a concrete calculation. Note that to construct a main ideal is equivalent to a rotation, this is similar to the normal subgroup.

---

[5]Note that the boundaries between distinct sextants in a lattice are similar to that in Fig. 5.4.

5.5.6 *Application to monolayer graphite*

Graphite has layer structure, and inside the layer there are quite strong interatomic interactions, and there are only weak interactions between the different layers. Therefore, a monolayer graphite was an important model for the study of graphite. But recently the monolayer graphite has become a real material due to the development of low dimensional technique. Now let us evaluate the elastic constants for the monolayer graphite.

5.5.6.1 *Calculation of pair potential*

The *ab initio* calculated total energy for a monolayer graphite sheet using the FLAPW method has been completed. Here, the Rose function is taken to fit the total energy curve as Rose[Ros84]

$$E(x) = -a[1 + \alpha(x-x_0)]e^{-\alpha(x-x_0)} \tag{5.92}$$

where x is the lattice constant, x_0 is the lattice constant at equilibrium state. The parameters are listed in Table 5.10. The calculated C-C pair potential can be obtained using Eq. (5.73). Now let us use Rose function to fit the calculated potential data, the fitting parameters are listed in Table 5.11. Note that Table 5.10 corresponds to Table 5.9 associated with three sources of total energy curves respectively.

5.5.6.2 *Calculated results of elastic constants*

For convenience, we define

$$e_{ij} = \frac{c}{2}C_{ij} \tag{5.93}$$

where c is lattice constant of bulk hcp crystal in c-axis direction, e_{ij} and C_{ij} are the ij-th elastic constant and the ij-th component of elastic modulus tensor respectively. Then e_{ij} can be expressed as

$$e_{ij} = \frac{1}{s_0}\frac{\partial^2 E}{\partial \epsilon_i \partial \epsilon_j}\Big|_{\epsilon_i=0, \epsilon_j=0} \tag{5.94}$$

Table 5.10 Parameters for fitting total energy of graphite sheet

	a(eV/atom)	$\alpha(A^{-1})$	$x_0(A)$
ab initio 1	7.41	3.038	1.429
ab initio 2	8.69	2.842	1.415
exp	7.39	3.029	1.421

Table 5.11 The calculated C-C potential with Rose cohesive energy

a (ev)	$\alpha(A^{-1})$	$r_0(A)$
3.456	3.095	1.594
3.795	2.900	1.622
3.422	3.083	1.589

where s_0 is the equilibrium atomic area in the plane as

$$s_0 = \frac{3\sqrt{3}}{4}x_0^2,$$

and $\check{\mathbf{E}}$ is deformation energy per atom in the monolayer.

The strains are given as follows.

$$\begin{cases} x' = x(1+\epsilon) & y' = y \quad \text{for } e_{11} \\ x' = x(1+\epsilon_1) & y' = y(1+\epsilon_2) \quad \text{for } e_{12} \\ x' = \dfrac{x}{\sqrt{1+\epsilon^2}} & y' = y + \dfrac{x\epsilon}{\sqrt{1+\epsilon^2}} \quad \text{for } e_{66} \end{cases} \quad (5.95)$$

According to a similar method to that discussed before, the deformation energy $\check{\mathbf{E}}(\epsilon_i, \epsilon_j)$ can be expressed as the sum of pair potentials, and thus the elastic constants can be expressed as

$$e_{11} = \frac{1}{24\sqrt{3}x_0} \sum_{\alpha \in E} \left\{ |\alpha| \phi'(|\alpha|x_0) + \sqrt{3}|\alpha| \phi'(\sqrt{3}|\alpha|x_0) \right\}$$
$$+ \frac{1}{8\sqrt{3}} \sum_{\alpha \in E} \left\{ |\alpha|^2 \phi''(|\alpha|x_0) + 3|\alpha|^2 \phi''(\sqrt{3}|\alpha|x_0) \right\} \quad (5.96)$$

and

$$e_{12} = e_{66} = \frac{-1}{24\sqrt{3}x_0} \sum_{\alpha \in E} \left\{ |\alpha| \phi'(|\alpha|x_0) + \sqrt{3}|\alpha| \phi'(\sqrt{3}|\alpha|x_0) \right\}$$
$$+ \frac{1}{24\sqrt{3}} \sum_{\alpha \in E} \left\{ |\alpha|^2 \phi''(|\alpha|x_0) + 3|\alpha|^2 \phi''(\sqrt{3}|\alpha|x_0) \right\} \quad (5.97)$$

where ϕ' and ϕ'' are the first and second derivatives regarding to ϕ respectively. Thus we have

$$e_{11} = \frac{1}{6\sqrt{3}}\check{\mathbf{E}}'(x_0) + \frac{1}{2\sqrt{3}}\check{\mathbf{E}}''(x_0) \quad (5.98)$$

Table 5.12 Comparison of calculated elastic constants and experimental values

	$e_{11}(N/m)$	$e_{12}(N/m)$	$e_{66}(N/m)$
Cal	315.9	105.3	105.3
Cal	324.2	108.1	108.1
Cal	313.2	104.5	104.5
Exp	378.6	94.5	—
Exp	355.1	60.3	147.4

and
$$e_{12} = e_{66} = \frac{-1}{6\sqrt{3}}\check{E}'(x_0) + \frac{1}{6\sqrt{3}}\check{E}''(x_0) \tag{5.99}$$
The equilibrium condition of a lattice is
$$E'(x_0) = 0.$$
Therefore,
$$e_{11} = 3e_{12} = 3e_{66} = \frac{1}{2\sqrt{3}}vE''(x_0).$$
The experimental values of e_{ij} are calculated from the relation $e_{ij} = \frac{c}{2}C_{ij}$, in which $c = 6.7A$[Che96; Wei82]. The calculated results are quite acceptable although many-body interactions have been not included.

5.5.7 Coordination number in a hexagonal lattice

Similar to the case in a square lattice, it can be proven that for an arbitrary integer n, the number of solutions of equation
$$x^2 + xy + y^2 = n \tag{5.100}$$
is equal to six times the difference of factors of n as $3k+1$ and that as $3k+2$. That is
$$N(n) = 6 \sum_{d|n} g(d) \tag{5.101}$$
where the arithmetic function $g(n)$ is
$$g(n) = \begin{cases} 1 & \text{when } n \equiv 1 (mod\ 3) \\ -1 & \text{when } n \equiv 2 (mod\ 3) \\ 0 & \text{when } 3|n \end{cases} \tag{5.102}$$

5.6 Summary

Chapter 5 presents the Möbius inverse formulas on the Gaussian and Eisenstein's integers with some applications, and includes the following features[Che96; She2003].

(1) For readers' convenience, the inverse square lattice cohesion problem was introduced at the beginning of Chapter 5 accompanied with the concept of the multiplicative semigroup. The method for 2D square lattice can be considered the inspiration for analyzing the 3D lattice inversion problem.

(2) This chapter also defined the irreducible Gaussian integers, then introduced the unique factorization of Gaussian integers and the sufficient and necessary condition of irreducibility, also defined the corresponding Möbius functions and established the sum rule, on the basis of which the Möbius inverse formula on Gaussian integers was determined.

(3) Similarly, the Möbius inverse formula on Eisenstein's integers was established.

(4) As an example of the Möbius inverse formula on Gaussian integers, the Bruns version of 2D AFT was developed with application to VLSI. As an example of the Möbius inverse formula on Eisenstein's integers, the interatomic potential in a single graphite sheet was introduced with application to elastic constant calculation.

(5) The coordination numbers or degeneracy problems for 2D lattices were discussed in detail.

Note that the modification of the sum rule with corresponding applications is left for future study.

Chapter 6

Inverse Lattice Problems

6.1 A Brief Historical Review

The study of crystal structures was initiated from the identification of cleavage plane orientations in the 1820s. Fifty years later, E.S. Fedoras (1853–1919) and A.M. Schröflies (1853–1928) proposed 230 space groups of crystallographic structures independently. In 1912, Max von Laue discovered the x-rays diffraction in crystals. In 1915, W.H. Bragg and W.L. Bragg established the method for the determination of crystal structures, and in 1919 P.P. Weald (1888–1985) proposed the concept of reciprocal lattices. All of these are related to three dimensional crystal structures, since the preparation of the free standing thin film specimen and atomic chain have emerged only in the last 20 years.

Chapter 5 discussed the inverse problems in atomic chains and monolayer. But a more challenging question is whether it will be feasible to extend the trick to problems that are not simply one- or two-dimensional. This chapter mainly presents the 3D inverse lattice problems.

Fig. 6.1 Left: Max von Laue (1879–1960). Right: W.H. Bragg (1862–1942) and W.L.Bragg (1890–1971).

6.2 3D Lattice Inversion Problem

It is well known that the average cohesive energy per atom $\mathbb{E}(x)$ can be expressed as the sum of all the interatomic potentials as

$$\mathbb{E}(x) = \frac{1}{2}\sum_{i\neq j}\Phi(r_{i,j}) + \frac{1}{6}\sum_{i\neq j\neq k\neq i}\Phi(r_{i,j}, r_{j,k}, r_{i,k}) + ... \quad (6.1)$$

where $\Phi(r_{i,j})$ represents the pair potential between the i-th atom and the j-th atom with distance $r_{i,j}$, and $\Phi(r_{i,j}, r_{j,k}, r_{i,k})$ the three body potential among atoms i, j, k, and so on. For most application purposes, pairwise potential terms in Eq. (6.1) play the leading role, and they are easily applied to various systems[In general, most angle dependent potentials are not available for molecular dynamics in a system with a huge number of atoms, and the EAM potentials are hard to obtain for most binary and ternary systems]. If only the pairwise terms are considered, it follows that

$$\mathbb{E}(x) = \frac{1}{2}\sum_{i\neq j}\Phi(r_{i,j}) = \frac{1}{2}\sum_{n=1}^{\infty}r(n)\Phi(b(n)x) \quad (6.2)$$

where x is the nearest neighbor distance. Regarding the reference atom located at origin, $r(n)$ and $b(n)$ represent the n-th coordination number and n-th neighbor distance respectively. Note that $b(1) = 1$ in this notation. For a general crystal structure, the sets $\{b(n)\}$ and $\{r(n)\}$ can be calculated by a simple computer program. In other words, $\{b(n)\}$ and $\{r(n)\}$ are known. Now the inverse lattice problem is how to extract the interatomic pair potentials from empirical or *ab initio* calculated data of $\mathbb{E}(x)$ based on equation Eq. (6.2).

6.2.1 *CGE solution*

In 1980, Carlsson, Gellat and Ehrenreich (CGE)[Car80; Esp80] proposed a method to extract the pair potentials from cohesive energy curve based on Eq. (6.2). Note that an *ab initio* band structure calculation can provide the total energy dependance $E_{tot}(x)$. Similar to Eq. (6.2), the cohesive energy per atom $\mathbb{E}(x) = E_{tot}(x) - E_{tot}(\infty)$ in CGE work is expressed as

$$\mathbb{E}(x) = \sum_{n=1}^{\infty}r_n\Phi(b_n x) \quad (6.3)$$

where r_n and b_n are the same to $r(n)$ and $b(n)$ in Eq. (6.2) respectively. Note that in the original CGE work, x denotes the lattice constant instead of the nearest neighbor distance. Here is a simplified treatment. Successively, Carlsson et al. define two operators as $\mathbb{T}_n$ and $\mathbb{T}$:

Definition 6.1.

$$\mathbb{T}_n \Phi(x) = r_n \Phi(b_n x) \quad \text{and} \quad \mathbb{T} = \mathbb{T}_1 + \sum_{n=2}^{\infty} \mathbb{T}_n \qquad (6.4)$$

then Eq. (6.3) becomes

$$\mathbb{E}(r) = \mathbb{T}\Phi(x) = [\mathbb{T}_1 + \sum_{n=2}^{\infty} \mathbb{T}_n] \Phi(x) \qquad (6.5)$$

Therefore, the formal solution of the inverse lattice problem can be expressed as

$$\Phi(x) = \mathbb{T}^{-1} \mathbb{E}(r) = [\mathbb{T}_1 + \sum_{n=2}^{\infty} \mathbb{T}_n]^{-1} \mathbb{E}(x) \qquad (6.6)$$

Note that

$$\mathbb{T}^{-1} = (1 + \mathbb{T}_1^{-1} \sum_{n=2}^{\infty} \mathbb{T}_n)^{-1} \mathbb{T}_1^{-1}$$

$$= \mathbb{T}_1^{-1} - \mathbb{T}_1^{-2} \sum_{n=2}^{\infty} \mathbb{T}_n + \mathbb{T}_1^{-3} \sum_{n_1,n_2=2}^{\infty} \mathbb{T}_{n_1} \mathbb{T}_{n_2}$$

$$- \mathbb{T}_1^{-4} \sum_{n_1,n_2,n_3=2}^{\infty} \mathbb{T}_{n_1} \mathbb{T}_{n_2} \mathbb{T}_{n_3} + \ldots \qquad (6.7)$$

Therefore,

$$\Phi(x) = (\frac{1}{r_1})\mathbb{E}(\frac{x}{b_1}) - \sum_{n=2}^{\infty} (\frac{r_n}{r_1^2}) \mathbb{E}(\frac{b_n x}{b_1^2}) + \sum_{p,q=2}^{\infty} (\frac{r_p r_q}{r_1^3}) \mathbb{E}(\frac{b_p b_q x}{b_1^3})$$

$$- \sum_{p,q,s=2}^{\infty} (\frac{r_p r_q r_s}{r_1^4}) \mathbb{E}(\frac{b_p b_q b_s x}{b_1^4}) + \ldots \qquad (6.8)$$

This is called CGE solution. Note that in the present notation $b_1 = 1$, (6.8) can be further simplified as

Theorem 6.1.

$$\Phi(x) = (\frac{1}{r_1})\mathbb{E}(x) - \sum_{n=2}^{\infty} (\frac{r_n}{r_1^2}) \mathbb{E}(b_n x) + \sum_{p,q=2}^{\infty} (\frac{r_p r_q}{r_1^3}) \mathbb{E}(b_p b_q x)$$

$$- \sum_{p,q,s=2}^{\infty} (\frac{r_p r_q r_s}{r_1^4}) \mathbb{E}(b_p b_q b_s x) + \ldots \qquad (6.9)$$

Equation(6.9) includes infinite number of summations, and each of them includes infinite terms. This caused time-consuming in calculation, and complicity in analysis.

6.2.2 Bazant iteration

In order to increase the convergence speed of the solution of Eq. (6.2), Bazant and Kaxiras proposed a numerical iteration technique [Baz96; Baz97].

Taking

$$\Phi_0(x) = \frac{1}{r_1}\mathbb{E}(x) \qquad (6.10)$$

as the initial condition which is the nearest neighbor approximation. They choose the iteration equation as

$$r_1\Phi_{k+1}(x) = \mathbb{E}(x) - \sum_{p=2}^{\infty} \mathbb{T}_p\Phi_k(x) \qquad (6.11)$$

The result of the iterative procedure should be

$$r_1\Phi(x) = \mathbb{E}(x) - \sum_{p=2}^{\infty} r_p\Phi(b_p x) \qquad (6.12)$$

or

$$\mathbb{E}(x) = \sum_{p=1}^{\infty} r_p\Phi(b_p x) \qquad (6.13)$$

The numerical solution can be obtained by Bazant iteration procedure Eq. (6.11) quickly. Note that Bazant iteration is also a convenient tool to extract three-body potentials.

6.3 Möbius Inversion for a General 3D Lattice

Now let us make a congruence analysis for square numbers and the sum of two squares. First, the multiplication modulo 8 or n^2 modulo 8 is as shown in Table 6.1. Second, denote S_2 as the sum of two square modulo 8, then $S_2 = \{0, 1, 2, 4, 5\}$ as in Table 6.2. By using the multiplication of two S_2, it can be checked that $S_2 \bigotimes S_2$ is congruent to S_2 modulo 8. In other words, the sum of two squares forms a closure regarding multiplications. For the case of simple cubic lattice, the corresponding S_3 is congruent to $\{0, 1, 2, 4, 5, 6\}$ as shown in Table 6.3).

Obviously, $S_3 \bigotimes S_3$ is not congruent to S_3. In other words, the distance function is not completely multiplicative for a general lattice. Or, there is no semigroup structure of lattice distances in a general 3D lattice. However, in the section on 2D square lattice in Chapter 5, it has been shown

Table 6.1 $n^2 \pmod 8$

n	0	1	2	3	4	5	6	7
$n^2 (\bmod\ 8)$	0	1	4	1	0	1	4	1

Table 6.2 $(m^2 + n^2)(\bmod\ 8)$

$\oplus$	0	1	4
0	0	1	4
1	1	2	4
4	4	5	1

Table 6.3 $(k^2 + m^2 + n^2)(\bmod\ 8)$

$\oplus$	0	1	2	4	5
0	0	1	2	4	5
1	1	2	3	5	6
4	4	5	6	0	1

that there must exist a Möbius inverse formula for any lattice, provided that the multiplicative closure is hold, no matter the concrete lattice is 2- or 3-dimensional.

Therefore, in order to solve a 3D inverse lattice problem by using the Möbius inversion method, the priority matter should examine the closeness of multiplicative operations of b(n) for the 3D lattice under consideration. For convenience, let us take a congruent analysis.

A congruent analysis of a simple cubic lattice for estimating the closeness of $b(n)$ (or $b^2(n)$) is shown in Table 6.1–Table 6.3, which include three priority issues: $n^2(\bmod\ 8)$, $\{m^2 + n^2\}(\bmod\ 8)$ and $\{k^2 + m^2 + n^2\}(\bmod\ 8)$. The result of $(k^2 + m^2 + n^2)$ modulo 8 include all the residues except 7. In other words, the sum of three squares may take all integers except $8k + 7$. Therefore, $b(n)$ can not be $\sqrt{7}, \sqrt{15}, \sqrt{23}, \sqrt{31}, \ldots$ for a simple cubic structure, and closeness does not hold, for example we can not find any $b(n)$ such that $b(n) = \sqrt{3}\sqrt{5} = b(3)b(5)$. In brief, the distance function $b(n)$ for a general lattice does not satisfy the multiplicative closure.

For a general case as

$$\mathbb{E}(x) = \frac{1}{2} \sum_{n=1}^{\infty} r(n)\Phi(b(n)x) \qquad (6.14)$$

which is exactly the same as Eqs. (6.2) or (6.12), we simply to work with the multiplicative closure of the set $\{b(n)\}$ to construct a multiplicative semigroup $\{B(n)\}$, which contains all the elements of $\{b(n)\}$ and all products of them. At the same time, the set of shell occupation numbers, $\{r(n)\}$ is extended to a set, $\{R(n)\}$, that assigns zero to all the elements $B(n) \notin \{b(n)\}$, which are called 'virtual lattice point shells'. Thus, it follows that

$$R(n) = \begin{cases} r(b^{-1}[B(n)]) & B(n) \in \{b(n)\} \\ 0 & B(n) \notin \{b(n)\} \end{cases} \tag{6.15}$$

and

$$\begin{aligned}\mathbb{E}(x) &= \frac{1}{2} \sum_{n=1}^{\infty} r(n) \Phi(b(n)x) \\ &= \frac{1}{2} \sum_{n=1}^{\infty} R(n) \Phi(B(n)x)\end{aligned} \tag{6.16}$$

Based on the multiplicative closeness of set $\{B(n)\}$, the inverse theorem for a general 3D lattice is given by

Theorem 6.2.

$$\mathbb{E}(x) = \frac{1}{2} \sum_{n=1}^{\infty} R(n) \Phi(B(n)x) \Rightarrow \Phi(x) = 2 \sum_{n=1}^{\infty} J(n) \mathbb{E}(B(n)x) \tag{6.17}$$

where the inversion coefficient $J(n)$ satisfies a modified sum rule as

$$\begin{cases} \sum_{B(m)B(m)=B(k)} J(n)R(m) = \delta_{k,1} \\ \text{or} \\ \sum_{B(n)|B(k)} J(n)R(B^{-1}\{\frac{B(k)}{B(n)}\}) = \delta_{k,1} \end{cases} \tag{6.18}$$

in which $B(n)|B(k)$ means that $B(n)$ runs over all the factors of $B(k)$ including $B(1)$ and $B(k)$, they satisfy $\frac{B(k)/B(n)}{\in}\{B(n)\}$, and the left part of Eq. (6.18) is also called a modified Dirichlet inverse. In fact, we have

$$2\sum_{n=1}^{\infty} J(n)\mathbb{E}(B(n)\ x)$$

$$= 2\sum_{n=1}^{\infty} J(n)[\frac{1}{2}\sum_{m=1}^{\infty} R(m)\Phi(B(n)B(m)\ x)]$$

$$= \sum_{k=1}^{\infty}[\sum_{B(n)B(m)=B(k)} J(n)R(m)]\Phi(B(k)\ x)$$

$$= \sum_{k=1}^{\infty}[\sum_{B(n)|B(k)} J(n)R(B^{-1}\frac{B(k)}{B(n)})]\Phi(B(k)\ x)$$

$$= \sum_{k=1}^{\infty} \delta(k,1)\Phi(B(k)\ x) = \Phi(x)$$

Obviously, if $b(2)-b(1) = b(2)-1 << 1$, i.e., the first two shells are very close each other, then according to the construction rule of $B(n)$ the significantly high density of supplementary virtual shells will appear, the convergence of the inversion become slower and worse. Note that convergence problems have not not been mentioned in the above discussion even though they are very important[Sza47]. One way to solve these convergence problems is to choose suitable fitting functions for calculated cohesive energy data. And after that, for fitting the converted data, we have to take care of the selecting of the functional form for interatomic potentials. Sometimes, a theoretically incompletely solved problem can be treated in a technical way which is accumulated partly by theory and partly by experience.

6.4 Inversion Formulas for some Common Lattice Structures

6.4.1 *Inversion formula for a fcc lattice*

In order to evaluate the inversion coefficients for a face-centered cubic structure, we shall write down the structure–dependent relation between cohesive energy and interatomic pair potentials. From Fig. 6.2(a), the cohesive

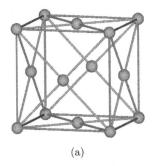
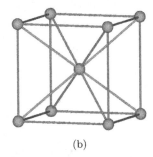

Fig. 6.2 (a) The face-centered structure and (b) body-centered structure

energy per atom in a fcc structure can be expressed as

$$\mathbb{E}(x) = \frac{1}{2} \sum_{m,n,\ell \neq 0,0,0} \Phi(\sqrt{2(m^2+n^2+\ell^2)}\, x)$$

$$+ \frac{3}{2} \sum_{m,n,\ell} \Phi(\sqrt{2[(m-\frac{1}{2})^2 + (n-\frac{1}{2})^2 + \ell^2]}\, x)$$

$$= 6\Phi(x) + 3\Phi(\sqrt{2}x) + 12\Phi(\sqrt{3}x) + 6\Phi(2x) + 12\Phi(\sqrt{5}x) + \cdots \quad (6.19)$$

where x is the nearest neighbor distance. Thus the concrete calculation for $b(n), r(n), B(n)$ and $R(n)$ can be obtained. Especially, $b(n)$ can be evaluated simply from congruence analysis. As mentioned above, $\{m^2 + n^2 + \ell^2\}(mod\ 8)$ runs over all the integers except $8k+7$, thus $\{m^2 + n^2 + \ell^2\}(mod\ 16)$ takes all integers but $16k+7$ and $16k+15$. Then we can see that, $2\{m^2 + n^2 + \ell^2\}(mod\ 16)$ runs over all evens but $16k+14$. On the other hand, $x \equiv 2\{(m-1/2)^2 + (n-1/2)^2 + (\ell-1/2)^2\}$ runs over all odd numbers. Therefore, it can be concluded that $[b(n)]^2$ runs over all integers but $16k+14$. Thus it is natural to expand $\{b(n)\}$ to $\{B(n)\}$ as

$$B(n) = \sqrt{n} \quad (6.20)$$

where $n=16k+14$ corresponds to virtual lattice points or virtual shell. Therefore we have

$$B(mn) = B(m)B(n) = \sqrt{mn} \quad (6.21)$$

and

$$B^{-1}\{\frac{B(k)}{B(n)}\} = \frac{k}{n} \quad (6.22)$$

Based on Eq. (6.15) the set of R(n) is obtained, and from Eq. (6.18) J(n) satisfies

$$\sum_{n|k} J(n)R(\frac{k}{n}) = \delta(k,1) \qquad (6.23)$$

Then the inversion coefficients can be extracted such as

$$J(1)R(1) = 1 \quad \text{or} \quad J(1) = \frac{1}{R(1)} = \frac{1}{12}$$

$$J(1)R(2) + J(2)R(1) = 0$$

$$\text{or} \quad J(2) = \frac{J(1)R(2)}{R(1)} = -\frac{(1/12) \times 6}{12} = \frac{-1}{24}$$

$$J(1)R(3) + J(3)R(1) = 0$$

$$\text{or} \quad J(3) = \frac{J(1)R(3)}{R(1)} = -\frac{(1/12) \times 24}{12} = \frac{-1}{6}$$

and in general

$$J(n) = -\frac{1}{R(1)} \sum_{d|n, d<n} J(d)R(\frac{n}{d}) \quad n > 1.$$

The calculated inversion coefficients for fcc structure are shown in Table 6.4, which is available for many fcc metals such as Al, Cu, Ni, Pd, Ag, Au.

According to the inversion coefficients, the concrete inversion formula can be expressed as

$$\Phi(x) = \frac{1}{6}\mathbb{E}(x) - \frac{1}{12}\mathbb{E}(\sqrt{2}x) - \frac{1}{3}\mathbb{E}(\sqrt{3}x) - \frac{1}{8}\mathbb{E}(2x) + ... \qquad (6.24)$$

The number of virtual lattice points is relative small($\sim 1/16$), the "anomaly" effect caused by virtual points does not cause any serious problem. This is why the approximations with a few neighbors are used very successful for metals with fcc structure. From the Table 6.4, the inversion coefficients do not have obvious increasing or decreasing tendency, thus the convergence of Eq. (6.24) mainly depends on the asymptotic behavior of cohesive energy curve at $x \to \infty$.

Table 6.4 Inversion coefficients for a fcc structure

n	1	2	3	4	5	6	7	8	9	10
$[B(n)]^2$	1	2	3	4	5	6	7	8	9	10
R(n)	12	6	24	12	24	8	48	6	36	24
J(n)	$\frac{1}{12}$	$-\frac{1}{24}$	$-\frac{1}{6}$	$-\frac{1}{16}$	$-\frac{1}{6}$	$\frac{1}{9}$	$-\frac{1}{3}$	$\frac{1}{32}$	$\frac{1}{12}$	0

n	11	12	13	14	15	16	17	18	19	20
$[B(n)]^2$	11	12	13	14	15	16	17	18	19	20
R(n)	24	24	72	0	48	12	48	30	72	24
J(n)	$-\frac{1}{6}$	$\frac{7}{72}$	$-\frac{1}{2}$	$\frac{1}{3}$	$\frac{1}{3}$	$-\frac{1}{64}$	$-\frac{1}{3}$	$-\frac{17}{72}$	$-\frac{1}{2}$	$\frac{5}{24}$

n	21	22	23	24	25	26	27	28	29	30
$[B(n)]^2$	21	22	23	24	25	26	27	28	29	30
R(n)	48	24	48	8	84	24	96	48	24	0
J(n)	1	0	$-\frac{1}{3}$	$-\frac{5}{72}$	$-\frac{1}{4}$	$\frac{1}{3}$	$-\frac{1}{3}$	$\frac{1}{12}$	$-\frac{1}{6}$	$\frac{2}{9}$

n	31	32	33	34	35	36	37	38	39	40
$[B(n)]^2$	31	32	33	34	35	36	37	38	39	40
R(n)	96	6	96	48	48	36	120	24	48	24
J(n)	$-\frac{2}{3}$	$\frac{1}{128}$	0	0	1	$\frac{61}{432}$	$-\frac{5}{6}$	$\frac{1}{3}$	$\frac{5}{3}$	$-\frac{1}{24}$

n	41	42	43	44	45	46	47	48	49	50
$[B(n)]^2$	41	42	43	44	45	46	47	48	49	50
R(n)	48	48	120	24	120	0	96	24	108	30
J(n)	$-\frac{1}{3}$	$-\frac{14}{9}$	$\frac{5}{6}$	$\frac{5}{24}$	$\frac{1}{2}$	$\frac{1}{3}$	$\frac{2}{3}$	$\frac{13}{288}$	$\frac{7}{12}$	$\frac{13}{24}$

n	51	52	53	54	55	56	57	58	59	60
$[B(n)]^2$	51	52	53	54	55	56	57	58	59	60
R(n)	120	0	144	12	48	48	168	48	96	48
J(n)	$-\frac{5}{6}$	$\frac{2}{3}$	$-\frac{5}{3}$	$\frac{1}{256}$	$\frac{5}{3}$	$\frac{2}{9}$	$\frac{7}{6}$	$\frac{5}{12}$	$\frac{2}{3}$	$\frac{2}{3}$

6.4.2 Inversion formula in a bcc structure

The relation between cohesive energy and interatomic potentials in a body-centered cubic lattice (see Fig. 6.2(b)) can be expressed as

$$\mathbb{E}(x) = \frac{1}{2} \sum_{m,n,\ell \neq 0,0,0} \Phi(\sqrt{\frac{4}{3}(m^2 + n^2 + \ell^2)}\, x)$$

$$+ \frac{1}{2} \sum_{m,n,\ell} \Phi(\sqrt{\frac{4}{3}[(m-1/2)^2 + (n-1/2)^2 + (\ell-1/2)^2]}\, x) \quad (6.25)$$

Table 6.5 Inversion coefficients for a bcc structure

n	1	2	3	4	5	6	7	8
$[B(n)]^2$	1	$\frac{4}{3}$	$\frac{16}{9}$	$\frac{64}{27}$	$\frac{8}{3}$	$\frac{256}{81}$	$\frac{32}{9}$	$\frac{11}{3}$
R(n)	8	6	0	0	12	0	0	24
J(n)	$\frac{1}{8}$	$-\frac{3}{32}$	$\frac{9}{128}$	$\frac{27}{512}$	$-\frac{3}{16}$	$\frac{81}{2048}$	$\frac{9}{32}$	$-\frac{3}{8}$

n	9	10	11	12	13	14	15	16
$[B(n)]^2$	4	$\frac{1024}{243}$	$\frac{128}{27}$	$\frac{44}{9}$	$\frac{16}{3}$	$\frac{4096}{729}$	$\frac{512}{81}$	$\frac{19}{3}$
R(n)	8	0	0	0	6	0	0	24
J(n)	$-\frac{1}{8}$	$-\frac{243}{8192}$	$\frac{81}{256}$	$\frac{9}{16}$	$\frac{3}{32}$	$\frac{729}{32768}$	$\frac{81}{256}$	$-\frac{3}{8}$

n	17	18	19	20	21	22	23	24
$[B(n)]^2$	$\frac{176}{27}$	$\frac{20}{3}$	$\frac{64}{9}$	$\frac{16384}{2187}$	8	$\frac{2048}{243}$	$\frac{76}{9}$	$\frac{704}{81}$
R(n)	0	24	0	0	24	0	0	0
J(n)	$-\frac{81}{128}$	$\frac{3}{8}$	$\frac{27}{128}$	$\frac{2187}{131072}$	$-\frac{3}{8}$	$\frac{1215}{4096}$	$\frac{9}{16}$	$\frac{81}{128}$

n	25	26	27	28	29	30	31	32
$[B(n)]^2$	$\frac{80}{9}$	9	$\frac{256}{27}$	$\frac{88}{9}$	$\frac{65536}{6561}$	$\frac{32}{3}$	$\frac{8192}{729}$	$\frac{304}{27}$
R(n)	0	32	0	0	0	12	0	0
J(n)	$\frac{9}{16}$	$-\frac{1}{2}$	$-\frac{297}{512}$	$\frac{9}{8}$	$\frac{6561}{524288}$	$\frac{3}{4}$	$\frac{2187}{8192}$	$-\frac{81}{128}$

n	33	34	35	36	37	38	39	40
$[B(n)]^2$	$\frac{2816}{243}$	$\frac{55}{3}$	$\frac{320}{27}$	12	$\frac{1024}{81}$	$\frac{352}{27}$	$\frac{262144}{19683}$	$\frac{40}{3}$
R(n)	0	48	0	30	0	0	0	24
J(n)	$-\frac{1215}{2048}$	$-\frac{3}{4}$	$-\frac{81}{128}$	$\frac{9}{32}$	$\frac{1863}{2048}$	$-\frac{81}{32}$	$-\frac{19683}{2097152}$	$-\frac{3}{8}$

n	41	42	43	44	45	46	47	48
$[B(n)]^2$	$\frac{121}{9}$	$\frac{128}{9}$	$\frac{43}{3}$	$\frac{44}{3}$	$\frac{32768}{2187}$	$\frac{1216}{81}$	$\frac{11264}{729}$	$\frac{140}{9}$
R(n)	0	0	24	24	0	0	0	0
J(n)	$\frac{9}{8}$	$-\frac{117}{128}$	$-\frac{3}{8}$	$\frac{3}{8}$	$\frac{15309}{65536}$	$\frac{81}{128}$	$\frac{2187}{4096}$	$\frac{9}{8}$

n	49	50	51	52	53	54	55	56
$[B(n)]^2$	$\frac{1280}{81}$	16	$\frac{4096}{243}$	$\frac{152}{9}$	17	$\frac{52}{3}$	$\frac{1408}{81}$	$\frac{1048576}{59049}$
R(n)	0	8	0	0	48	24	0	0
J(n)	$\frac{81}{128}$	$-\frac{9}{64}$	$-\frac{9477}{8192}$	$\frac{9}{8}$	$-\frac{3}{4}$	$-\frac{3}{8}$	$\frac{243}{64}$	$\frac{59049}{8388608}$

Note that $\mathbb{E}(x) = \frac{1}{2}\sum_{n=1}^{\infty} R(n)\Phi(B(n)\,x)$, The $R(n)$, $B(n)$ and $I(n)$ for a bcc structure can be calculated as shown in Table 6.5. Noted that a congruence analysis shows that $[b(n)]^2 = \frac{4}{3}(m^2 + n^2 + \ell^2)$ corresponds to $4k/3$ without $k \equiv 7 \pmod 8$, and $[b(n)]^2 = \frac{4}{3}[(m-1/2)^2 + (n-1/2)^2 + (\ell-1/2)^2]$ corresponds to $(4k+3)/3$. The multiplications of them produce a lot of terms as $4k/3^n$, $(4k+1)/3^n$ and $(4k+3)/3^n$. Therefore, there are many virtual lattice points produced by inversion in a bcc structure. This analysis is useful for checking the calculated table.

Therefore, the cohesion energy per atom in bcc structure can be expressed as

$$\mathbb{E}(x) = 4\Phi(x) + 3\Phi(\sqrt{\tfrac{4}{3}}x) + 6\Phi(\sqrt{\tfrac{8}{3}}x)$$
$$+ 12\Phi(\sqrt{\tfrac{11}{3}}x) + 4\Phi(2x) + 3\Phi(\tfrac{\sqrt{16}}{3}x) + \ldots \qquad (6.26)$$

and the pair potential can be expressed as

$$\Phi(x) = \frac{1}{4}\mathbb{E}(x) - \frac{3}{16}\mathbb{E}(\sqrt{\tfrac{4}{3}}x) + \frac{9}{64}\mathbb{E}(\tfrac{4}{3}x) - \frac{27}{256}\mathbb{E}(\sqrt{\tfrac{64}{27}}x) - \frac{3}{8}\mathbb{E}(\sqrt{\tfrac{8}{3}}x)$$
$$+ \frac{81}{1024}\mathbb{E}(\sqrt{\tfrac{256}{81}}x) + \frac{9}{16}\mathbb{E}(\sqrt{\tfrac{32}{9}}x) - \frac{3}{4}\mathbb{E}(\sqrt{\tfrac{11}{3}}x)$$
$$- \frac{1}{4}\mathbb{E}(2x) - \frac{243}{4096}\mathbb{E}(\sqrt{\tfrac{1024}{243}}x) - \frac{81}{128}\mathbb{E}(\sqrt{\tfrac{128}{27}}x)$$
$$+ \frac{9}{8}\mathbb{E}(\sqrt{\tfrac{44}{9}}x) + \frac{3}{16}\mathbb{E}(\sqrt{\tfrac{16}{3}}x) + \ldots \qquad (6.27)$$

Comparing the above two equations, the terms of

$$\frac{9}{64}\mathbb{E}(\tfrac{4}{3}x),\ -\frac{27}{256}\mathbb{E}(\sqrt{\tfrac{64}{27}}x),\ \frac{81}{1024}\mathbb{E}(\sqrt{\tfrac{256}{81}}x),\ \frac{9}{16}\mathbb{E}(\sqrt{\tfrac{32}{9}}x),$$
$$\frac{243}{4096}\mathbb{E}(\sqrt{\tfrac{1024}{243}}x),\ \frac{81}{128}\mathbb{E}(\sqrt{\tfrac{128}{27}}x),\ \frac{9}{8}\mathbb{E}(\sqrt{\tfrac{44}{9}}x),\ldots$$

are "abnormal" terms caused by vanished $R(n)$, they emerge like the ghost lines in optics. From the first 100 inversion coefficients $R(n)$ it is seen that there are 45 real atomic shells accompanied with 55 virtual shells. Note that most of the 55 virtual shells correspond to nonzero $J(n)$. Therefore, it is not unexpected that the convergence of inversion formula is not very good, and the instability appears quite often in numerically calculated results of the bcc structure materials. This can be understood, since for a bcc structure we have $b(2) - b(1) = 2/\sqrt 3 \simeq 0.1547 \ll 1$, there are already 9 zeros in the first $15R(n)$. On the other hand, sometimes later it is possible to design a way to improve the bcc materials by adjusting the $b(n) - R(n)$ distribution with impurity.

6.4.3 Inversion formula for the cross potentials in a $L1_2$ structure

In a binary intermetallic compound Ni_3Al with the $L1_2$ structure (see Fig. 6.3(a)), we denote the nearest neighboring distance as x, then the average cohesive energy per atom is expressed as

$$\mathbb{E}^{Ni_3Al}(x) = \frac{1}{4}\mathbb{E}^{Al-Al}(x) + \frac{3}{4}\mathbb{E}^{Ni-Ni}(x) + \frac{1}{4}\mathbb{E}^{Al-Ni}(x) \qquad (6.28)$$

where the interatomic potentials Φ_{Al-Al} between identical Al atoms can be calculated by inversion of cohesive energies of fcc Al, and Φ_{Ni-Ni} can be obtained by that of fcc Ni respectively. Therefore these partial cohesive energies involved in Ni_3Al can be expressed as

$$\mathbb{E}^{Al-Al}(x) = \frac{1}{2}\sum_{(m,n,\ell)\neq(0,0,0)} \Phi_{Al-Al}(\sqrt{2(m^2+n^2+\ell^2)}x), \qquad (6.29)$$

$$\mathbb{E}^{Ni-Ni}(x) = \frac{1}{2}[\sum_{m,n,\ell} \Phi_{Ni-Ni}(\sqrt{2\{(m-1/2)^2+(n-1/2)^2+\ell^2\}}x)$$
$$+ \sum_{(m,n,\ell)\neq(0,0,0)} \Phi_{Ni-Ni}(\sqrt{2(m^2+n^2+\ell^2)}x)], \qquad (6.30)$$

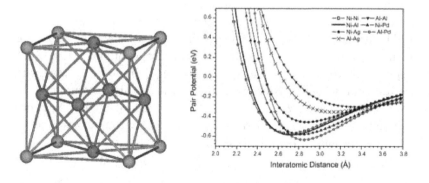

Fig. 6.3 $L1_2$ structure, and interatomic potentials in Ni_3Al.

and
$$\mathbb{E}^{Al-Ni}(x) = 3 \sum_{m,n,\ell} \Phi_{Al-Ni}(\sqrt{2([m-1/2]^2 + [n-1/2]^2 + \ell^2)}x) \quad (6.31)$$

respectively. Now let us consider how to extract the cross potentials $\Phi_{Al\text{-}Ni}$ from the partially cohesive energy $\mathbb{E}^{Al\text{-}Ni}(x)$. Note that there is no factor of $1/2$ in Eq. (6.31) since there is no repeated contribution from the interaction between Al and Ni. In order to evaluate the corresponding $b(n)$ and $r(n)$, only Eq. (6.31) is needed. Since $[b(n)]^2 = 2([m-1/2]^2 + [n-1/2]^2 + \ell^2) = 2(m^2 + n^2 + \ell^2) - 2(m+n) + 1$ runs over all the odd numbers which form a semigroup under multiplication, thus $B(n) = b(n), R(n) = r(n)$ and we have

$$\mathbb{E}^{Al-Ni}(x) = 3 \sum_{n=1}^{\infty} r(n) \Phi_{Al-Ni}(b(n)x) \quad (6.32)$$

and

$$\Phi_{Al-Ni}(x) = \frac{1}{3} \sum_{n=1}^{\infty} I(n) \mathbb{E}^{Al-Ni}(b(n)x) \quad (6.33)$$

with

$$\sum_{b(m)b(n)=b(k)} I(n)r(m) = \delta_{k,1} \quad (6.34)$$

The calculated inversion coefficients of $I(n)$ are listed in Table 6.6. The Morse parameters for $\Phi_{Al-Al}(x), \Phi_{Ni-Ni}(x)$, and $\Phi_{Al-Ni}(x)$ are listed in Table 6.7. The corresponding interatomic potentials are shown in Fig. 6.3(b). The site preference and phonon density of states can demonstrated on Table 6.8 and Fig. 6.4 respectively.

Now let us present an example for calculating the interatomic potentials in Ni_3Al with $L1_2$ structure. In general, one may calculate the potentials between identical atoms from the *ab initio* calculation of bulk Al and bulk Ni respectively first, then the $\Phi_{Al\text{-}Al}(x)$ and $\Phi_{Ni\text{-}Ni}(x)$ are extracted with lattice inversion formula. Now according to the sublattice structures the partial cohesive energies $E_{Al\text{-}Al}(x)$ and $E_{Ni\text{-}Ni}(x)$ are calculated. Next one can evaluate $\mathbb{E}^{Al\text{-}Ni}(x)$, and extract $\Phi_{Al\text{-}Ni}(x)$. Table 6.8 shows a set of resultant interatomic potential curves fit by Morse function

$$\Phi(r) = D_0\{\exp[-2\alpha(R - R_0)] - 2exp[-\alpha(R - R_0)]\} \quad (6.35)$$

Kang *et al.* have calculated the site preference of third element in Ni_3Al by using the lattice inverted pair potentials, the results are in good agreement with experiment[Kan2002].

Table 6.6 The inversion coefficients in $L1_2$ structure

n	1	2	3	4	5	6	7	8	9	10
$[b(n)]^2$	1	3	5	7	9	11	13	15	17	19
r(n)	12	24	24	48	36	24	72	48	48	72
I(n)	$\frac{1}{12}$	$-\frac{1}{6}$	$-\frac{1}{6}$	$-\frac{1}{3}$	$\frac{1}{12}$	$-\frac{1}{6}$	$\frac{1}{2}$	$\frac{1}{3}$	$\frac{1}{3}$	$-\frac{1}{2}$

n	11	12	13	14	15	16	17	18	19	20
$[b(n)]^2$	21	23	25	27	29	31	33	35	37	39
r(n)	48	48	84	96	24	96	96	48	120	48
I(n)	1	$-\frac{1}{3}$	$-\frac{1}{6}$	$-\frac{1}{4}$	$-\frac{1}{3}$	$-\frac{1}{6}$	$-\frac{2}{3}$	0	1	$-\frac{5}{6}$

n	21	22	23	24	25	26	27	28	29	30
$[b(n)]^2$	41	43	45	47	49	51	53	55	57	59
r(n)	48	120	120	96	108	48	72	144	96	72
I(n)	$-\frac{1}{3}$	$-\frac{5}{6}$	$-\frac{1}{2}$	$-\frac{2}{3}$	$\frac{7}{12}$	1	$-\frac{1}{2}$	$-\frac{1}{3}$	$\frac{3}{4}$	$-\frac{1}{2}$

n	31	32	33	34	35	36	37	38	39	40
$[b(n)]^2$	61	63	65	67	69	71	73	75	77	79
r(n)	120	144	48	168	96	48	192	120	96	96
I(n)	$-\frac{5}{6}$	$-\frac{5}{3}$	$\frac{5}{3}$	$-\frac{7}{6}$	$\frac{2}{3}$	$-\frac{1}{3}$	$-\frac{4}{3}$	$\frac{5}{6}$	$\frac{2}{3}$	$-\frac{2}{3}$

n	41	42	43	44	45	46	47	48	49	50
$[b(n)]^2$	81	83	85	87	89	91	93	95	97	99
r(n)	48	120	120	96	108	48	72	144	96	72
I(n)	1	$-\frac{5}{6}$	$\frac{1}{3}$	$\frac{1}{3}$	$-\frac{2}{3}$	3	$\frac{5}{3}$	$\frac{5}{3}$	$-\frac{5}{3}$	$\frac{5}{6}$

n	51	52	53	54	55	56	57	58	59	60
$[b(n)]^2$	101	103	105	107	109	111	113	115	117	119
r(n)	72	240	96	72	120	144	96	240	216	96
I(n)	$-\frac{1}{2}$	$-\frac{5}{3}$	$-\frac{10}{3}$	$-\frac{1}{2}$	$-\frac{5}{6}$	$\frac{7}{3}$	$-\frac{2}{3}$	$-\frac{1}{3}$	$\frac{19}{63}$	2

6.5 Atomistic Analysis of the Field-Ion Microscopy Image of Fe_3Al

The study of the field-ion microscopy (FIM) shows perfect ring structure in FIM images for pure metals. The size of concentric rings corresponds to the local curvature radius of the specimen's tip surface. Larger planar density of atoms corresponds to more prominent poles in the FIM images[Mul69; Mil89].

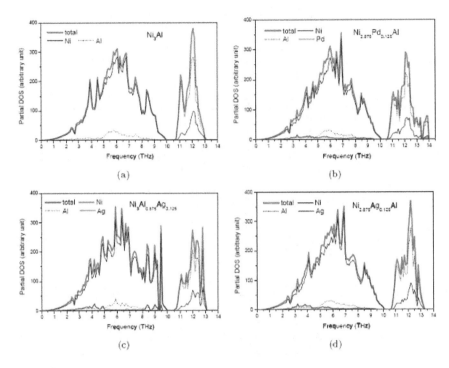

Fig. 6.4 Phonon DOS in Ni_3Al with foreign atoms: (a) Pure Ni_3Al, (b) with Pd replacing part of Ni, (c) with Ag replacing part of Al, (d) with Ag replacing part of Ni.

Table 6.7 The Morse parameters for $\Phi_{Al-Al}(x)$, $\Phi_{Ni-Ni}(x)$ and $\Phi_{Al-Ni}(x)$

Morse	$R_0(A)$	D_0(eV)	α
Al-Al	3.3855	0.2864	1.0637
Ni-Ni	2.7930	0.4364	1.3979
Al-Ni	2.7212	0.5859	1.6835

The clear rings also occur in FIM images for binary ordered alloys. However, there exists an additional phenomenon in experiment, which is called invisibility. That is, one of the species seems to have disappeared from the FIM images of binary alloys. Apparently, the invisibility can be explained by selective evaporation assuming that the applied voltage can induce evaporation of surface atoms from the tip sample. Conventionally, atoms with lower sublimation energy (in the pure metal state) are invisible

Table 6.8 Site preference of foreign atom in Ni_3Al

Ternary elements	$Coh_{Al} - -Coh_{Ni}$ (eV/atom)	Calculations	Experiments and other work
Ag	−0.0122	Al	Al
Si	−0.1276	Al	Al
Mg	−0.1390	Al	Al
Zr	−0.0899	Al	Al
Hf	−0.1457	Al	Al
Zn	−0.0794	Al	Al
W	−0.0191	Al	no/Al
Mo	−0.0042	no	no/Al
Sc	−0.1102	Al	Ni/Al
Pt	−0.0615	Ni	Ni
Co	−0.0893	Ni	Ni
Cu	−0.0234	Ni	Ni
Pd	−0.0736	Ni	Ni
Mn	−0.0605	Ni	no/Ni
Fe	−0.0762	Ni	no
Cr	−0.0350	Ni	no/Al

since the atoms with higher sublimation energy would preferably remain at the tip surface and construct the stable FIM image. This can explain the invisibility of Co in PtCo, Ni in Ni_4Mo, but it can not explain the case in Fe_3Al due to the change in bonding during the formation of an alloy.

Bonding energies of surface atoms of intermetallic compounds are calculated in order to predict the selective field evaporation behavior.

Now let us evaluate several inversion coefficients related to DO_3 structure such as Fe_3Al (see Fig. 6.5). The total cohesive energy of an average atom in Fe_3Al includes three partial energies

$$\mathbb{E}_{total}^{Fe_3Al}(x) = \mathbb{E}_{Al-Al}^{Fe_3Al}(x) + \mathbb{E}^{Fe\text{-}Fe}(x) + \mathbb{E}^{Fe\text{-}Al}(x) \qquad (6.36)$$

where x is the nearest neighbor distance. The partial cohesive energy $E^{Fe\text{-}Al}(x)$ is equal to[Che98]

$$E^{Fe\text{-}Al}(x) = \frac{1}{4} \sum_{m,n,\ell} \Phi_{Fe\text{-}Al}(\sqrt{(m-\frac{1}{2})^2 + (n-\frac{1}{2})^2 + (\ell-\frac{1}{2})^2} \frac{2x}{\sqrt{3}}) +$$

$$+ \frac{1}{4} \sum_{m,n,\ell} \Phi_{Fe\text{-}Al}(\sqrt{(m-\frac{1}{2})^2 + (n-\frac{1}{2})^2 + (\ell-\frac{1}{2})^2} \frac{4x}{\sqrt{3}}) +$$

$$+ \frac{1}{4} \sum_{m,n,\ell} \Phi_{Fe\text{-}Al}(\sqrt{m^2 + n^2 + (\ell-\frac{1}{2})^2} \frac{4x}{\sqrt{3}}) \qquad (6.37)$$

Based on these, one can obtain $b(n), r(n), B(n), R(n)$ and $J(n)$ for a DO_3 structure as in Table 6.9. Table 6.10 shows the experimental cohesive energies for several pure metals.

The calculated bonding energy at different surfaces for Fe_3Al are given in Table 6.11. Although the sublimation energy of aluminum is smaller than that of iron, the bonding energy E_b of Al atoms in Fe_3Al is larger than that of Fe atoms. We find that this is caused by the interaction with the third nearest neighbors. All of the third nearest neighbors of an Al atom in Fe_3Al are Al atoms, while at the corresponding distance Al $--$ Al is significantly larger than Fe $--$ Fe and Fe $--$ Al. On all the surfaces considered, the bonding energies Es of Al atoms are approximately $0.1--0.5eV$ higher than those of Fe atoms. Consequently, Fe atoms will be removed selectively from the tip surface and become invisible. This agrees with the experimental observation, see references in [Che98; Ge99].

In convention, one compares the sublimation energy value of pure metals Fe and Al to conclude that Al should be the invisible species[Kra87]. The present work has an opposite result since the reliable cross interactions between distinct atoms are involved. The similar calculation has been done for other intermetallic compounds[Ge99].

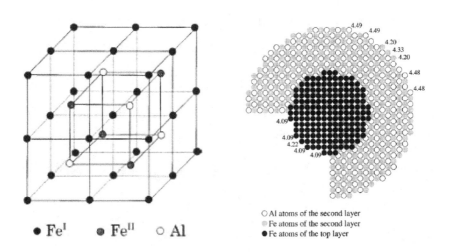

Fig. 6.5 Left: DO_3 structure of Fe_3Al. Right: A top view of a DO_3-type Fe_3Al tip.

Table 6.9 The inversion coefficients in DO$_3$ structure

n	1	2	3	4	5	6	7	8
$[B(n)]^2$	1	$\dfrac{4}{3}$	$\dfrac{16}{9}$	$\dfrac{64}{27}$	$\dfrac{256}{81}$	$\dfrac{11}{3}$	4	$\dfrac{1024}{243}$
R(n)	8	6	0	0	0	24	8	0
J(n)	$\dfrac{1}{8}$	$-\dfrac{3}{32}$	$\dfrac{9}{128}$	$-\dfrac{27}{512}$	$\dfrac{81}{2048}$	$-\dfrac{3}{8}$	$-\dfrac{1}{8}$	$\dfrac{-243}{8192}$

n	9	10	11	12	13	14	15	16
$[B(n)]^2$	$\dfrac{44}{9}$	$\dfrac{16}{3}$	$\dfrac{4096}{729}$	$\dfrac{19}{3}$	$\dfrac{176}{27}$	$\dfrac{20}{3}$	$\dfrac{64}{9}$	$\dfrac{16384}{2187}$
R(n)	0	0	0	24	0	24	0	0
J(n)	$\dfrac{9}{16}$	$\dfrac{3}{16}$	$\dfrac{729}{32768}$	$-\dfrac{3}{8}$	$-\dfrac{81}{128}$	$-\dfrac{3}{8}$	$-\dfrac{27}{128}$	$\dfrac{2187}{131072}$

n	17	18	19	20	21	22	23	24
$[B(n)]^2$	$\dfrac{76}{9}$	$\dfrac{704}{81}$	$\dfrac{80}{9}$	9	$\dfrac{256}{27}$	$\dfrac{65536}{6561}$	$\dfrac{304}{27}$	$\dfrac{2816}{243}$
R(n)	0	0	0	32	0	0	0	0
J(n)	$\dfrac{9}{16}$	$\dfrac{81}{128}$	$\dfrac{9}{16}$	$-\dfrac{1}{2}$	$\dfrac{27}{128}$	$\dfrac{6561}{524288}$	$-\dfrac{81}{128}$	$\dfrac{1215}{2048}$

n	25	26	27	28	29	30	31	32
$[B(n)]^2$	$\dfrac{35}{3}$	$\dfrac{320}{27}$	12	$\dfrac{1024}{81}$	$\dfrac{262144}{19683}$	$\dfrac{121}{9}$	$\dfrac{43}{3}$	44/3
R(n)	48	0	30	0	0	0	24	24
J(n)	$-\dfrac{3}{4}$	$-\dfrac{81}{128}$	9/32	$-\dfrac{405}{2048}$	$-\dfrac{19683}{2097152}$	$\dfrac{9}{8}$	$-\dfrac{3}{8}$	$\dfrac{3}{8}$

n	33	34	35	36	37	38	39	40
$[B(n)]^2$	$\dfrac{1216}{81}$	$\dfrac{11264}{729}$	$\dfrac{140}{9}$	$\dfrac{1280}{81}$	16	$\dfrac{4096}{243}$	17	$\dfrac{52}{3}$
R(n)	0	0	0	0	0	0	48	24
J(n)	$\dfrac{81}{128}$	$\dfrac{2187}{4096}$	$\dfrac{9}{8}$	$\dfrac{81}{128}$	$-\dfrac{1}{64}$	$\dfrac{729}{4096}$	$-\dfrac{3}{4}$	$-\dfrac{3}{8}$

n	41	42	43	44	45	46	47	48
$[B(n)]^2$	$\dfrac{1048576}{59049}$	$\dfrac{484}{27}$	$\dfrac{172}{9}$	$\dfrac{176}{9}$	$\dfrac{59}{3}$	$\dfrac{4864}{243}$	$\dfrac{45056}{2187}$	$\dfrac{560}{27}$
R(n)	0	0	0	0	72	0	0	0
J(n)	$\dfrac{59049}{8388608}$	$-\dfrac{81}{32}$	$\dfrac{9}{16}$	$-\dfrac{9}{8}$	$\dfrac{9}{8}$	$\dfrac{1215}{2048}$	$\dfrac{15309}{32768}$	$-\dfrac{81}{64}$

n	49	50	51	52	53	54	55	56
$[B(n)]^2$	$\dfrac{5120}{243}$	$\dfrac{64}{3}$	$\dfrac{67}{3}$	$\dfrac{16384}{729}$	$\dfrac{68}{3}$	$\dfrac{208}{9}$	$\dfrac{209}{9}$	$\dfrac{4194304}{177147}$
R(n)	0	0	24	0	48	0	0	0
J(n)	$-\dfrac{1215}{2048}$	$\dfrac{117}{512}$	$-\dfrac{3}{8}$	$\dfrac{5103}{32768}$	$\dfrac{3}{8}$	$\dfrac{9}{16}$	$\dfrac{9}{4}$	$-\dfrac{177147}{33554432}$

6.6 Interaction between Unlike Atoms in B_1 and B_3 structures

6.6.1 *Expression based on a cubic crystal cell*

Assuming there is an ordered binary alloy with a B_1 structure (see Fig. 6.6(a)), the minima distance between unlike atoms is $a/2$, where a is

Table 6.10 Sublimation energy of different pure metals

Element	Pt	Co	Ni	Mo	Fe	Al
Sublimation(eV/atom)	5.84	4.39	4.44	6.82	4.28	3.39

Table 6.11 Bonding energy of different surface atoms in Fe_3Al

Bonding energy(eV)	(100)	(110)	(111)
Fe^I	5.319	5.381	4.317/4.648
Fe^{II}	5.322	5.618	4.595
Al	5.822	5.723	5.145

lattice constant. Then the set of distance function relative to $a/2$ includes

$$b(n) = \begin{cases} \sqrt{(i+1/2)^2 + j^2 + k^2} \\ \sqrt{i^2 + (j+1/2)^2 + k^2} \\ \sqrt{k^2 + j^2 + (k+1/2)^2} \\ \sqrt{(i+1/2)^2 + (j+1/2)^2 + (k+1/2)^2} \end{cases} \quad (6.38)$$

where i, j, k are integers.

Similarly let us assume that there is an ordered binary alloy with a B_3 structure (see Fig. 6.6(b)), the minima distance between unlike atoms is

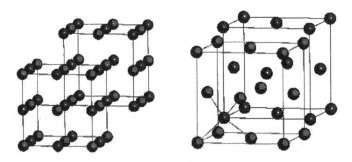

Fig. 6.6 Left: B_1 divide into two fcc structures. Right: B_3 into two fcc.

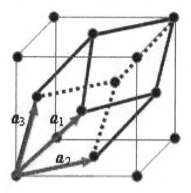

Fig. 6.7 Unit cell of fcc structure

$\sqrt{3}a/4$. Then the set of distance function relative to $\sqrt{3}a/4$ includes

$$b(n) = \begin{cases} \sqrt{(i+1/4)^2 + (j+1/4)^2 + (k+1/4)^2} \\ \sqrt{(i+3/4)^2 + (j+3/4)^2 + (k+1/4)^2} \\ \sqrt{(i+3/4)^2 + (j+1/4)^2 + (k+3/4)^2} \\ \sqrt{(i+1/4)^2 + (j+3/4)^2 + (k+3/4)^2} \end{cases} \quad (6.39)$$

6.6.2 Expression based on a unit cell

For a fcc structure, its unit cell is shown as Fig. 6.7. The corresponding unit vectors in the unit cell are

$$\begin{cases} \mathbf{a}_1 = \dfrac{a}{2}(\mathbf{x}+\mathbf{y}) \\ \mathbf{a}_2 = \dfrac{a}{2}(\mathbf{z}+\mathbf{x}) \\ \mathbf{a}_3 = \dfrac{a}{2}(\mathbf{x}+\mathbf{y}) \end{cases} \quad (6.40)$$

where $\mathbf{x}, \mathbf{y}, \mathbf{z}$ are the unit vectors in the Descartes coordination. Obviously, an arbitrary lattice vector in a fcc structure can be expressed as

$$R = m\mathbf{a}_1 + n\mathbf{a}_2 + \ell\mathbf{a}_3. \quad (6.41)$$

Now let us consider the B1 structure, the relative shift between the two fcc sublattices can be considered as a half of diagonal of the cubic as

$$\mathbf{r}_a = \frac{a}{2}\mathbf{x} + \frac{a}{2}\mathbf{y} + \frac{a}{2}\mathbf{z} \quad (6.42)$$

Table 6.12 Inversion coefficients in B_1 structure

n	1	2	3	4	5	6	7	8	9	10
$[B(n)]^2$	1	3	5	9	11	13	15	17	19	21
R(n)	6	8	24	30	24	24	0	48	24	48
J(n)	1/6	-2/9	-2/3	-29/54	-2/3	-2/3	16/9	-4/3	-2/3	-4/3
n	11	12	13	14	15	16	17	18	19	20
$[B(n)]^2$	25	27	29	33	35	37	39	41	43	45
R(n)	30	32	72	48	48	24	0	96	24	72
J(n)	11/6	76/81	-2	4/9	-4/3	-2/3	16/9	-8/3	-2/3	10/9
n	21	22	23	24	25	26	27	28	29	30
$[B(n)]^2$	49	51	53	55	57	59	61	63	65	67
R(n)	54	48	72	0	48	72	72	0	96	24
J(n)	-3/2	20/9	-2	16/3	4/9	-2	-2	32/9	8/3	-2/3
n	31	32	33	34	35	36	37	38	39	40
$[B(n)]^2$	69	73	75	77	81	83	85	87	89	91
R(n)	96	48	56	96	102	72	48	0	144	48
J(n)	-8/3	-4/3	-10	-8/3	-52/243	-2	28/3	16/3	-4	-4/3
n	41	42	43	44	45	46	47	48	49	50
$[B(n)]^2$	93	95	97	99	101	105	107	109	111	113
R(n)	48	0	48	72	168	96	72	72	0	96
J(n)	-4/3	16/3	-4/3	14/3	-14/3	-104/9	-2	-2	16/9	-8/3
n	51	52	53	54	55	56	57	58	59	60
$[B(n)]^2$	115	117	121	123	125	129	131	133	135	137
R(n)	48	120	78	48	144	144	120	48	0	96
J(n)	-4/3	-2/9	1/2	52/9	-8	-20/9	-10/3	-4/3	-640/81	-8/3

then it is given by

$$\mathbf{R} - \mathbf{r}_a = m\frac{a}{2}(\mathbf{x}+\mathbf{y}) + n\frac{a}{2}(\mathbf{z}+\mathbf{x}) + \ell\frac{a}{2}(\mathbf{y}+\mathbf{z}) - (\frac{a}{2}\mathbf{x} + \frac{a}{2}\mathbf{y} + \frac{a}{2}\mathbf{z})$$

$$= \frac{a}{2}x(m+n-1) + \frac{a}{2}y(m+\ell-1) + \frac{a}{2}z(n+\ell-1) \quad (6.43)$$

the corresponding distance function relative to $a/2$ is

$$b(n) = \sqrt{(m+\ell-1)^2 + (m+n-1)^2 + (n+\ell-1)^2} \quad (6.44)$$

and

$$E_{+-}^{B1}(a) = \sum_{m,n,\ell} \Phi(\sqrt{(m+\ell)^2 + (m+n-1)^2 + (n+\ell)^2}\frac{a}{2}) \quad (6.45)$$

The corresponding crossing inversion coefficients for a B_1 structure are shown in Table 6.12.

Now let us consider the B_3 structure, the relative shift between the two fcc sublattices can be considered as a quarter of diagonal of the cubic as

$$\mathbf{r}'_a = \frac{a}{4}\mathbf{x} + \frac{a}{4}\mathbf{y} + \frac{a}{4}\mathbf{z} \quad (6.46)$$

then it is given by

$$\mathbf{R} - \mathbf{r}'_a = m\frac{a}{2}(\mathbf{x}+\mathbf{y}) + n\frac{a}{2}(\mathbf{z}+\mathbf{x}) + \ell\frac{a}{2}(\mathbf{y}+\mathbf{z}) - (\frac{a}{4}\mathbf{x} + \frac{a}{4}\mathbf{y} + \frac{a}{4}\mathbf{z})$$

$$= \frac{a}{2}x(m+n-1/2) + \frac{a}{2}y(m+\ell-1/2) + \frac{a}{2}z(n+\ell-1/2) \qquad (6.47)$$

the corresponding distance function relative to $a/2$ is

$$b(n) = \sqrt{(m+\ell-1/2)^2 + (m+n-1/2)^2 + (n+\ell-1/2)^2} \qquad (6.48)$$

and

$$E_{+-}^{B3}(a)$$
$$= \sum_{m,n,\ell} \Phi(\sqrt{(m+\ell-1/2)^2 + (m+n-1/2)^2 + (n+\ell-1/2)^2}\frac{a}{2}) \qquad (6.49)$$

The corresponding crossing inversion coefficients for a B_3 structure are shown in Table 6.13.

6.7 The Stability and Phase Transition in NaCl

Note that the above two kinds of ordered alloy in last section are hypothetical structures, the real binary systems with B_1 or B_3 structures are most likely ionic compounds.

Most alkali-chlorides appear with B_1 structure, in which NaCl is the most famous one. In order to extract the crossing interionic potential $Phi_{+-}(x)$, we need to calculate the partial cohesive energy $E_{+-}^{B1}(a)$, but it is not possible in practice. However, we are able to both the total cohesive energies $E_{tot}^{B1}(a)$ and $E_{tot}^{B3}(a)$, and

$$E_{tot}^{B1}(a) = E_{\text{Na-Na}}^{fcc} + E_{\text{Cl-Cl}}^{fcc} + E_{\text{Na-Cl}}^{B1}(a) \qquad (6.50)$$

and

$$E_{tot}^{B3}(a) = E_{\text{Na-Cl}}^{B3}(a) + E_{\text{Na-Na}}^{fcc} + E_{\text{Cl-Cl}}^{fcc} \qquad (6.51)$$

Taking the subtraction of the two cohesive energies it is obtained

$$\Delta E_{\text{Na-Cl}}(a) = E_{tot}^{B3}(a) - E_{tot}^{B1}(a) = E_{\text{Na-Cl}}^{B3}(a) - E_{\text{Na-Cl}}^{B1}(a) \qquad (6.52)$$

This indicates that the difference of two cohesive energies, $E_{tot}^{B1}(a) - E_{tot}^{B3}(a)$, is only dependent on the interaction between the cross terms $Na^+ - Cl^-$. Note that the interionic potential consists of the long-range part and the short-range part. For example,

$$E_{\text{Na-Cl}}(a) = E_{\text{Na-Cl}}^{Cou}(a) + E_{\text{Na-Cl}}^{SR}(a) \qquad (6.53)$$

Table 6.13 Cross inversion coefficients in B_3 structure

n	1	2	3	4	5	6	7	8	9	10
$[B(n)]^2$	1	$\frac{11}{3}$	$\frac{19}{3}$	9	$\frac{35}{3}$	$\frac{121}{9}$	$\frac{43}{3}$	17	$\frac{59}{3}$	$\frac{67}{3}$
R(n)	4	12	12	16	24	0	12	24	36	12
J(n)	$\frac{1}{4}$	$-\frac{3}{4}$	$-\frac{3}{4}$	-1	$-\frac{3}{2}$	$\frac{9}{4}$	$-\frac{3}{4}$	$-\frac{3}{2}$	$\frac{9}{4}$	$-\frac{3}{4}$

n	11	12	13	14	15	16	17	18	19	20
$[B(n)]^2$	$\frac{209}{9}$	25	$\frac{83}{3}$	$\frac{91}{3}$	33	$\frac{107}{3}$	$\frac{115}{3}$	$\frac{361}{9}$	41	$\frac{385}{9}$
R(n)	0	28	36	24	36	36	24	0	24	0
J(n)	$\frac{9}{2}$	$-\frac{7}{4}$	$-\frac{9}{4}$	$\frac{3}{2}$	$\frac{15}{4}$	$-\frac{9}{4}$	$\frac{3}{2}$	$\frac{9}{4}$	$-\frac{3}{2}$	9

n	21	22	23	24	25	26	27	28	29	30
$[B(n)]^2$	$\frac{131}{3}$	$\frac{139}{3}$	49	$\frac{1331}{27}$	$\frac{155}{3}$	$\frac{473}{9}$	$\frac{163}{3}$	57	$\frac{179}{3}$	$\frac{187}{3}$
R(n)	60	36	28	0	48	0	12	60	60	24
J(n)	$-\frac{15}{4}$	$-\frac{9}{4}$	$-\frac{7}{4}$	$-\frac{27}{4}$	-3	$\frac{9}{2}$	$-\frac{3}{4}$	$\frac{9}{4}$	$-\frac{15}{4}$	$\frac{15}{2}$

n	31	32	33	34	35	36	37	38	39	40
$[B(n)]^2$	65	$\frac{203}{3}$	$\frac{211}{3}$	$\frac{649}{9}$	73	$\frac{665}{9}$	$\frac{227}{3}$	$\frac{235}{3}$	81	$\frac{737}{9}$
R(n)	48	48	36	0	48	0	60	24	52	0
J(n)	-3	-3	$-\frac{9}{4}$	$\frac{27}{2}$	-3	9	$-\frac{15}{4}$	$\frac{3}{2}$	$\frac{3}{4}$	$\frac{9}{2}$

n	41	42	43	44	45	46	47	48	49	50
$[B(n)]^2$	$\frac{251}{3}$	$\frac{2299}{27}$	$\frac{259}{3}$	89	$\frac{817}{9}$	$\frac{275}{3}$	$\frac{283}{3}$	97	$\frac{299}{3}$	$\frac{913}{9}$
R(n)	84	0	48	24	0	60	36	48	96	0
J(n)	$-\frac{21}{4}$	$-\frac{81}{4}$	-3	$-\frac{3}{2}$	$\frac{9}{2}$	$\frac{27}{4}$	$-\frac{9}{4}$	-3	-6	$\frac{27}{2}$

n	51	52	53	54	55	56	57	58	59	60
$[B(n)]^2$	$\frac{307}{3}$	105	$\frac{323}{3}$	$\frac{331}{3}$	$\frac{1001}{9}$	113	$\frac{347}{3}$	$\frac{355}{3}$	121	$\frac{371}{3}$
R(n)	36	72	48	36	0	72	60	48	52	96
J(n)	$-\frac{9}{4}$	$-\frac{15}{2}$	6	$-\frac{9}{4}$	9	$-\frac{9}{2}$	$-\frac{15}{4}$	-3	$-\frac{67}{4}$	-6

and

$$\Phi_{\text{Na-Cl}}(x) = \Phi^{Cou}_{\text{Na-Cl}}(x) + \Phi^{SR}_{\text{Na-Cl}}(x) \qquad (6.54)$$

The contribution of Coulomb terms can be treated by Ewald summation technique[Ewa21], while the short-range contribution by the Möbius inversion technique as follows.

For a B1-type ionic NaCl crystal, the relation between the partial cohesive energy $E^{B1}_{\text{Na-Cl}(SR)}$ and the crossing interionic potentials $\Phi^{SR}_{\text{Na-Cl}}$ is given by

$$\Delta E^{SR}_{\text{Na-Cl}}(a) = E^{B3(SR)}_{Na-Cl}(a) - E^{B1(SR)}_{Na-Cl}(a)$$

$$= \sum_{m,n,\ell} \Phi(\sqrt{(m+\ell-1/2)^2 + (m+n-1/2)^2 + (n+\ell-1/2)^2}\frac{a}{2})$$

$$- \sum_{m,n,\ell} \Phi(\sqrt{(m+\ell-1)^2 + (m+n-1)^2 + (n+\ell-1)^2}\frac{a}{2})$$

$$= \sum_{m,n,\ell} \Phi(\sqrt{\frac{4\{(m+\ell-1/2)^2 + (m+n-1/2)^2 + (n+\ell-1/2)^2\}}{3}}\frac{\sqrt{3}a}{4})$$

$$- \sum_{m,n,\ell} \Phi(\sqrt{\frac{4\{(m+\ell-1)^2 + (m+n-1)^2 + (n+\ell-1)^2\}}{3}}\frac{\sqrt{3}a}{4})$$

Note that here we take the lattice constant a as variable of cohesive energy expression for both B_1 and B_3 structures. Now we take the nearest neighbor distance $x = \sqrt{3}a/4$ as the variable for energy difference of the mixed virtual lattice $B_3 - B_1$, it follows that

$$\Delta E^{SR}_{\text{Na-Cl}}(x)$$

$$= \sum_{m,n,\ell} \Phi(\sqrt{\frac{4\{(m+\ell-1/2)^2 + (m+n-1/2)^2 + (n+\ell-1/2)^2\}}{3}}x)$$

$$- \sum_{m,n,\ell} \Phi(\sqrt{\frac{4\{(m+\ell-1)^2 + (m+n-1)^2 + (n+\ell-1)^2\}}{3}}x) \quad (6.55)$$

Computing and counting the like terms $\Phi(b(n)x)$, subsequently transferring the sets $\{b(n)\}$ and $\{r(n)\}$ to $\{B(n)\}$ and $\{R(n)\}$[1], it is given by

$$\Delta E^{SR}_{\text{Na-Cl}}(x) = \frac{1}{2}\sum_{n=1}^{\infty} R(n)\Phi^{SR}_{\text{Na-Cl}}(B(n)x) \quad (6.56)$$

Based on the inversion formula, it follows that

$$\Phi^{SR}_{\text{Na-Cl}}(x) = 2\sum_{n=1}^{\infty} J(n)\Delta E^{SR}_{\text{Na-Cl}}\{B(n)x\} \quad (6.57)$$

with J(n) determined by

$$\sum_{B(n)|B(k)} J(n)R[B^{-1}(\frac{B(k)}{B(n)})] = \delta_{k,1} \quad (6.58)$$

---

[1] In convention, one defines $\Delta \hat{E}^{SR}_{\text{Na-Cl}}(x)$ in order to keep $r(1) = R(1) > 0$

Table 6.14 Cross inversion coefficients in B1 structure accompanied by B3 structure

n	1	2	3	4	5	6	7	8	9	10
$[B(n)]^2$	1	$\frac{4}{3}$	$\frac{16}{9}$	$\frac{64}{27}$	$\frac{256}{81}$	$\frac{11}{3}$	4	$\frac{1024}{243}$	$\frac{44}{9}$	$\frac{16}{3}$
R(n)	4	−6	0	0	0	12	−8	0	0	0
J(n)	$\frac{1}{4}$	$-\frac{3}{8}$	$\frac{9}{16}$	$\frac{27}{32}$	$\frac{81}{64}$	$-\frac{3}{4}$	$\frac{1}{2}$	$\frac{243}{128}$	$-\frac{9}{4}$	$\frac{3}{2}$
n	11	12	13	14	15	16	17	18	19	20
$[B(n)]^2$	$\frac{4096}{729}$	$\frac{194}{3}$	$\frac{176}{27}$	$\frac{20}{3}$	$\frac{64}{9}$	$\frac{16384}{2187}$	$\frac{76}{9}$	$\frac{704}{81}$	$\frac{80}{9}$	9
R(n)	0	12	0	-24	0	0	0	0	0	9
J(n)	$\frac{729}{256}$	$-\frac{3}{4}$	$-\frac{81}{16}$	$\frac{3}{2}$	$\frac{27}{8}$	$\frac{2187}{512}$	$-\frac{9}{4}$	$-\frac{81}{8}$	$\frac{9}{2}$	−1

Combining the short-range term and the Coulomb term, the total cross pair potential in NaCl is given by

$$\Phi_{\text{Na-Cl}}(r) = D_{+-}\exp\{\gamma_{+-}(1-\frac{r}{R_{+-}})\} + \frac{q_+q_-}{4\pi\varepsilon_0 r} \quad (6.59)$$

The fitting parameters $D_{+-}, \gamma_{+-}, R_{+-}$ of the cross interatomic potential for NaCl are shown in Table 6.14 . Similarly, one can extract the interionic pair potentials $\Phi_{Na-Na}(r)$ and $\Phi_{Cl-Cl}(r)$ respectively. By using these potentials, some interesting simulations have been done. Note that the calculated short-range interaction for $Na^+ - Na^+$ is very small and neglected. The results have been applied to stability of B_1–NaCl, the disorder-order transition, high-pressure $B_1 - -B_2$ transition[Zha2002; Zha2003a]. The similar calculations have been done also for other B_1 structures such as LiCl, KCl, RbCl[Zha2003a; Zha2003b; Zha2003c] and CaO[Wan2005].

Note that the fitting function forms for different interatomic potentials are different. In practice, the function form for each calculated result is chosen to minimize the mean-square error. According to calculation, the function form of Φ_{--} takes the Morse function for NaCl, KCl, RbCl, the repel-exponential function for LiCl; the form of Φ_{+-} takes the repel-exponential function for all the four; the calculated Φ_{--} can be neglected except RbCl.

Obviously, the presence of apparent abnormal inversion coefficients shows

Table 6.15 Fitting parameters for short-range interactions in LiCl, NaCl, KCl, RbCl

Crystal	Ion pair	Function form	D(eV)	R(A)	γ
LiCl	++	—	—	—	—
	+−	Rep-Exp	1.3751	1.8750	5.9266
	−	Rep-Exp	0.4466	2.7929	8.7348
NaCl	++	—	—	—	—
	+−	Rep-Exp	0.2848	2.6499	8.6729
	−	Morse	0.0244	3.7338	11.3902
KCl	++	—	—	—	—
	+−	Rep-Exp	1.7149	2.3383	6.4557
	−	Morse	0.1177	3.7066	8.8093
RbCl	++	Rep-Exp	0.0572	2.6860	8.7701
	+−	Rep-Exp	0.5099	2.9085	8.2691
	−	Morse	0.0870	3.9386	7.8972

something interesting. Noted that when the electronic charge distribution is approximately spherical, the pair potentials can be easily accepted. That is the situation for both typical metals and ionic crystals. Now people may ask how about the materials with covalent bonding, such as semiconductor compounds. By using the multiple lattice inversion method similar to that for ionic crystal, Cai et al. have obtained a lot of interatomic potentials in which the main part come from the lattice inversion in various systems such as III-V semiconductors XY(X=Al,Ga, Y=N, P, As, Sb) and their pseudo binary alloys, as well SiC, ZnO, ZnS. In the static simulation based on inverted potentials, the calculated lattice constant, bulk modulo and cohesive energy based on inverted potentials are in good agreement with *ab initio* calculations. This indicates the self-consistency of the lattice inversion. When a small three-body term is added, the Cauchy deviation and phonon dispersion can be well explained. The results have been applied to high pressure phase transitions such as B_3-B_1 in GaAs[Cai2005; Cai2007a]. The lattice inversion has been also applied to AlN, GaN and InN with B_4 structure[Zha2005]. Note that the explanation of phase transition such as B_3-B_8 in AlAs, B_4-B_1 in AlN, GaN and InN need a great number of *ab initio* calculations[Cai2007b; Cai2007c].

6.8 Inversion of Stretching Curve

This section shows that the interatomic potentials can be converted approximately by using cohesive energy curve in a stretching process as

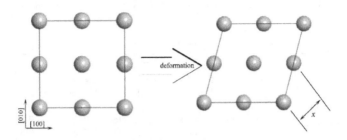

Fig. 6.8 Stretch deformation

shown in Fig. 6.8. Now let us consider another interesting example to apply the semi-group technique to studying a bcc crystal by using the stretcher strain curve along [110] direction. Denote the crystal constant as a, the interval between layers along [110] as x. Then the relation between E, the average cohesive energy for each atom during stretching process, and the interatomic potential $\Phi(x)$ can be expressed as

$$E(x) = \sum_{m=1}^{\infty} \sum_{n_1,n_2=-\infty}^{\infty} \Phi\left(\sqrt{m^2 x^2 + (n_1 + \frac{m}{2})^2 a^2 + (n_2 + \frac{m}{2})^2 \frac{a^2}{2}}\right)$$

$$= \sum_{m=1}^{\infty} \sum_{n_1,n_2=-\infty}^{\infty} \Phi\left(\sqrt{m^2 x^2 + (8n_1^2 + 8n_1 m + 4n_2^2 + 4n_2 m + 3m^2)\frac{a^2}{8}}\right)$$

$$= \sum_{m,n} h_{m,n} \Phi\left(\sqrt{m^2 x^2 + n\frac{a^2}{8} + \frac{3a^2}{8}}\right) \quad (6.60)$$

Here the coefficient $h_{m,n}$ is the number of solutions of the equation

$$n = 8n_1^2 + 8n_1 m + 4n_2^2 + 4n_2 m + 3m^2 - 3 = 2(2n_1+m)^2 + (2n_2+m)^2 - 3 \quad (6.61)$$

Define a function

$$\Psi(x) = \Phi\left(\sqrt{x^2 + \frac{3a^2}{8}}\right) \quad (6.62)$$

and a operator

$$A_{m,n}(x) = \sqrt{x^2 + \frac{3a^2}{8}} \quad (6.63)$$

Thus Eq. (6.60) can be expressed as

$$E(x) = \sum_{m,n} h_{m,n} \Psi\left(\sqrt{m^2 x^2 + n\frac{a^2}{8}}\right)$$

$$= \sum_{m,n} h_{m,n} \Psi(A_{m,n}(x)) \quad (6.64)$$

Note that there is semigroup relation

$$A_{m_1,n_1} \circ A_{m_2,n_2} = A_{m_1 m_2, m_1^2 n_2 + n_1} \tag{6.65}$$

The corresponding isomorphic matrix representation can be chosen as

$$\begin{bmatrix} m_1^2 & n_1 \\ 0 & 1 \end{bmatrix} \begin{bmatrix} m_2^2 & n_2 \\ 0 & 1 \end{bmatrix} = \begin{bmatrix} m_1^2 m_2^2 & m_1^2 n_2 + n_1 \\ 0 & 1 \end{bmatrix} \tag{6.66}$$

Therefore, from Eq. (6.60) it is given that

$$\Psi(x) = \sum_{m,n} g_{m,n} E(A_{m,n}(x)) \tag{6.67}$$

or

$$\Phi(x) = \sum_{m,n} g_{m,n} E\left(\sqrt{m^2\left(x^2 - \frac{3a^2}{8}\right) + n\frac{a^2}{8}}\right) \tag{6.68}$$

where $g_{m,n}$ satisfies

$$\sum_{A_1 \circ A_2 = A} g_{A_1} h_{A_2} = \delta_{A,I} \tag{6.69}$$

Note that the multiplicative relation in (6.65) is a noncommutative one such that

$$A_{m_2,n_2} \circ A_{m_1,n_1} = A_{m_1 m_2, m_2^2 n_1 + n_2} \neq A_{m_1 m_2, m_1^2 n_2 + n_1} \tag{6.70}$$

But this does not affect the calculation in Eq. (6.69) since $m_1^2 n_2 + n_1 = 0$ is equivalent to $m_2^2 n_1 + n_2 = 0$ when $m_1^2 m_2^2 = 1$. Of course, the resulting potential is different from that based on common cohesive energy curve. But the present section shows a different way to use semigroup in crystal physics.

6.9 Lattice Inversion Technique for Embedded Atom Method

It is well known that the pair potentials are not available for describing a number of complex properties, such as Cauchy discrepancy, vacancy formation energy, stability of a elemental material with bcc or diamond structures. Recognizing the requirement for many-body interatomic potentials for metals and alloys, the embedded-atom method (EAM) based on the density-functional theory proposed by Daw and Baskes in 1983[Daw83; Daw84] has proven to be a significant method for simulating the energetics of metals. The key point of EAM is to separate the cohesive energy into two parts: one mainly from the screened electro-static interaction between two ionic cores, the other from the kinetic energy and the exchange-correlation energy. The former contribution can be represented by the lattice sums of two-body potentials, while the latter contribution is represented by the so-called embedding function, which means the interaction of the immersed atom with the surrounding electron cloud. The embedding function is assumed to be a function of the 'local electron density ρ', which is a rigid superposition of atomic electron charge densities from all the other environmental atoms. The basic equations of EAM are

$$\begin{cases} \mathbb{E}_{coh} = \sum_i F(\rho_i) + \sum_i \Psi_i \\ \rho_i = \sum_{j \neq i} f_j(R_{ji}) \\ \Psi_i = \frac{1}{2} \sum_{j \neq i} \Phi_{ji}(R_{ji}) \end{cases} \quad (6.71)$$

For a single atom crystal structure, the cohesion of each atom can be expressed as

$$\begin{cases} E_{coh} = F(\rho) + \frac{1}{2} \sum_j \Phi(R_j) \\ \rho = \sum_j f(R_j) \end{cases} \quad (6.72)$$

Note that there are two similar summations as

$$\begin{cases} E_{coh}(x) - F[\rho(x)] = \frac{1}{2} \sum_{n=1}^{\infty} r(n) \Phi(b(n)x) \\ \rho(x) = \sum_{n=1}^{\infty} r(n) f(b(n)x) \end{cases} \quad (6.73)$$

where x represents the nearest neighbor distance, $b(n)$ and $r(n)$ are the fractional distance and the coordination number respectively. Correspondingly, we have two lattice inversion relations as[Xie94-1; Xie94-2; Xie95; Zha98]

$$\begin{cases} \Phi(x) = 2 \sum_{n=1}^{\infty} J(n) \{E^u(B(n)x) - F[\rho(B(n)x)]\} \\ f(x) = \sum_{n=1}^{\infty} J(n) \rho(B(n)x) \end{cases} \quad (6.74)$$

with the common coefficient J(n) satisfying

$$\sum_{B(n)|B(k)} J(n) R(B^{-1} \frac{B(k)}{B(n)}) = \delta_{k1} \quad (6.75)$$

where multiplicative semigroup $\{B(n)\}$ is extension from $\{b(n)\}$, the corresponding $\{R(n)\}$ is from $\{r(n)\}$ as before. Note that $f(x)$ is the effective charge density of an atom under spherical approximation, and in principle, the attempt for extracting the charge density of a single atom in solid is impossible. $\rho(x)$ can be calculated by using *ab initio* method with the model in which all the atoms positioned in sites of the complete lattice except removing one atom in the center of the crystal.

Baskes proposed the form for an embedded function as

$$F(\rho) = -F_0[1 - \ln\frac{\rho}{\rho_e}](\frac{\rho}{\rho_e}) \tag{6.76}$$

The cohesive energy term can be taken from Rose's universal equation[Ros84], that is

$$E^u(R) = -E_c(1 + a*)e^{-a*} \tag{6.77}$$

where

$$a* = \alpha(R/r_e - 1) \text{ and } \alpha = \sqrt{9B\Omega/E_c} \tag{6.78}$$

Here $E^u(R)$ is the universal function for a uniform expansion or contraction in the reference structure, B is the bulk modulus and Ω is the equilibrium atomic volume.

For a fcc structure, it is given by

$$\mathfrak{E}(R) = E^u(R) - F[\rho(R)]$$
$$= 6\Phi(R) + 3\Phi(\sqrt{2}\,R) + 12\Phi(\sqrt{3}\,R) + 6\Phi(2R) + 12\Phi(\sqrt{5}\,R) + \ldots \tag{6.79}$$

Similar to Eq. (6.24), the lattice inversion result can be expressed as

$$\Phi(x) = \frac{1}{6}\mathfrak{E}(R) - \frac{1}{12}\mathfrak{E}(\sqrt{2}\,R) - \frac{1}{3}\mathfrak{E}(\sqrt{3}\,R)$$
$$- \frac{1}{8}\mathfrak{E}(2R) - \frac{1}{3}\mathfrak{E}(\sqrt{5}\,R) + \ldots \tag{6.80}$$

For a bcc structure, it is given by

$$\mathfrak{E}(R) = E^u(R) - F[\rho(R)]$$
$$= 4\Phi(R) + 3\Phi(\sqrt{\frac{4}{3}}\,R) + 6\Phi(\sqrt{\frac{8}{3}}\,R) + 12\Phi(\sqrt{\frac{11}{3}}\,R) + \ldots \tag{6.81}$$

Similar to Eq. (6.26), the corresponding lattice-inversion result is

$$\Phi(x) = \frac{1}{4}\mathfrak{E}(R) - \frac{3}{16}\mathfrak{E}(\sqrt{\frac{4}{3}}\,R) + \frac{9}{64}\mathfrak{E}(\frac{4}{3}\,R) - \frac{27}{256}\mathfrak{E}(\sqrt{\frac{64}{27}}R)$$
$$- \frac{3}{8}\mathfrak{E}(\sqrt{\frac{8}{3}}\,R) + \frac{81}{1024}\mathfrak{E}(\sqrt{\frac{256}{81}}\,R)$$
$$+ \frac{9}{16}\mathfrak{E}(\sqrt{\frac{32}{9}}\,R) - \frac{3}{4}\mathfrak{E}(\sqrt{\frac{11}{3}}\,R)\ldots \tag{6.82}$$

By using the lattice inversion technique, the number of empirical parameters is reduced significantly. However, from above, we see the complexity of the lattice inversion for a bcc structure[Bas2000]. In fact, the 'strange' things in a bcc structure take place not only due to emergence of virtual lattices or ghost lattice points. A more serious problem is that the conventional EAM potential for a bcc structure always leads to lower bonding energy (or higher cohesive energy). This is attributed to that only one state is involved in the old version EAM without angle-dependence. Therefore, Laskowski proposed a double lattice inversion technique in 2000[Las2000] to overcome this shortcoming, different from the original EAM only based on the equilibrium state equation. Later, Baskes started the work on a multistate modified embedded atom method(MS-MEAM), in which the reference states are not only the the object phase, but also some metastable phases and transformation path. In MS-MEAM, all the traditional empirical parameters are reduced, and the variation in bonding energies with highly symmetric local arrangement of atoms implies the the angular dependence of the bonding[Bas2007].

Similarly, we can discuss the case of a binary intermetallic compounds as follows[Xie94-1; Xie94-2].

$$\begin{cases} E_{coh}(x) - F[\rho_A(x)] - F[\rho_B(x)] = \frac{1}{2} \sum_{n=1}^{\infty} r(n)\Phi(b(n)x) \\ \sum_{n=1}^{\infty} r(n)\Phi(b(n)x) = \sum_{n=1}^{\infty} r_{AA}(n)\Phi_{AA}(b(n)x) \\ \qquad + 2\sum_{n=1}^{\infty} r_{AB}(n)\Phi_{AB}(b(n)x) + \sum_{n=1}^{\infty} r_{BB}(n)\Phi(b_{BB}(n)x) \\ \rho_A(x) = \sum_{n=1}^{\infty} r_{AA}(n)f(b_{AA}(n)x) + \sum_{n=1}^{\infty} r_{AB}(n)f(b_{AB}(n)x) \\ \rho_B(x) = \sum_{n=1}^{\infty} r_{BA}(n)f(b_{AA}(n)x) + \sum_{n=1}^{\infty} r_{BB}(n)f(b_{BB}(n)x) \end{cases} \qquad (6.83)$$

where ρ_A is the 'local charge density' for atom A, namely the sum of charge densities from environmental atoms centered on an A atom. This can be calculated by using a rigid crystal lattice without the above A atom. Similar treatment can be given for ρ_B. $\Phi_{AA}(x)$ and $\Phi_{BB}(x)$ may be extracted from pure element metal calculation.

Now the formulation has been accomplished to extract EAM functions directly from calculations for first-principles energy vs atomic volume. The large number of calculated complex properties for Cu is in reasonable agreement with experiments. Applications of this formulation to multicomponent systems should be expected[Bas2007]. Also, it can be expected that the application of lattice inversion technique based on semigroup analysis to MS-MEAM will be further developed.

6.10 Interatomic Potentials between Atoms across Interface

6.10.1 *Interface between two matched rectangular lattices*

At an interface, the periodicity of atomic arrangement terminates. This termination means the loss of atomic neighbors at a free surface, and therefore the change of electron distribution. The structure of a specific interface may be very complex. This chapter proceeds from a ideal interface model, in which either side of the interface keeps the atomic arrangement as in a bulk crystal, considering the above termination and the adjustment of lattice scale for coherent matching. Based on these ideal interfacial system models, a series of *ab initio* calculation of total energies can be accomplished.

The inverse interfacial adhesion problem is to extract the interatomic potentials across the interface based on the *ab initio* interfacial adhesive energy curve. There are some premise in this stage: each side of the interface is a complete semi-infinite crystal structure without interface reconstruction during the changing of interface distance; the lattices of two sides are matched each other. In other words, we are going to discuss the coherent interface with two sides as semi-infinite lattice structure. Assume that the interface system consists of two parts, both of them are pure metals, with simple cubic structures. For simplification, we introduce a simple coherent interface model. The part one is a semi-infinite simple cubic lattice with lattice constant a, the transverse lattice constant of part two is due to the coherent property of the interface, the longitudinal lattice constant is b. Then the adhesive energy of the interfacial system can be expressed by interfacial potentials and the interface distance x as follows.

$$E(x) = \sum_{\ell_1,\ell_2=0}^{\infty} \sum_{n=-\infty}^{\infty} \Phi(\sqrt{(x+\ell_1 a+\ell_2 b)^2 + n^2 a^2}) \qquad (6.84)$$

In order to solve Eq. (6.84) easily, we introduce a function $\Delta(x)$ as

$$\Delta(x) = \sum_{n=-\infty}^{\infty} \Phi(\sqrt{x^2 + n^2 a^2}) \qquad (6.85)$$

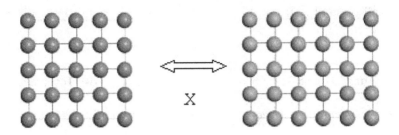

Fig. 6.9 Simple interface model

then Eq. (6.84) becomes

$$E(x) = \sum_{\ell_1,\ell_2=0}^{\infty} \Delta(x + \ell_1 a + \ell_2 b) \tag{6.86}$$

From Eq. (6.85) it follows that

$$\Delta(x) = E(x) - E(x+a) - E(x+b) + E(x+a+b) \tag{6.87}$$

Also, from Eq. (6.85) we have

$$\Delta(x) = \sum_{m=0}^{\infty} r(m)\Phi(A_m\{x\}) \tag{6.88}$$

where the action of the operator A_m to x is defined as

$$A_m\{x\} = \sqrt{x^2 + ma^2} \tag{6.89}$$

thus the operator A_m is a additive Cesàro operator, i.e.,

$$A_{m_1} A_{m_2} = A_{m_1+m_2} \tag{6.90}$$

In addition, $r(m)$ in Eq. (6.88) is

$$r(m) = \begin{cases} 1 & \text{if } m = 0 \\ 2 & \text{if } m = n^2 > 0 \\ 0 & \text{if } m \neq n^2 > 0 \end{cases} \tag{6.91}$$

By applying the additive Cesàro theorem to Eq. (6.88), we have

$$\Phi(x) = \sum_{m=0} I(m)\Delta(A_m\{x\}) \tag{6.92}$$

where the inversion coefficients I(m) satisfy

$$\sum_{0 \leq m \leq k} I(m) r(k-m) = \delta(k,0) \tag{6.93}$$

Finally,

$$\Phi(x) = \sum_{m=0}^{\infty} I(m)\{E(\sqrt{x^2+ma^2}) - E(\sqrt{x^2+ma^2}+a) - \\ - E(\sqrt{x^2+ma^2}+b) + E(\sqrt{x^2+ma^2}+a+b)\} \tag{6.94}$$

6.10.2 Metal/MgO interface

Now let us consider an inverse problem on an interface between face-centered cubic structure and MgO, or M(001)/MgO(001). Now there are many metal atoms (M) in one side, and many ions Mg^{++} and O^{--} in the other side. We start from a model with coherent lattices on interface as shown in Fig. 6.10. There are two structures under consideration since two potentials (across interface) $\Phi_{M-Mg^{++}}$ and $\Phi_{M-O^{--}}$ are to be extracted. In one structure each metal atom in interface is positioned right on the top of ion O^{--}, and in another case, the metal atom is on the top of ion Mg^{++}. From this the relation between adhesive energy and potentials can be expressed as

$$E_0 = \sum_{\ell,\ell'=0}^{\infty} \sum_{m,n=-\infty}^{\infty} \{\Phi_{M-O}(\sqrt{(x+\ell a+\ell' a)^2+(ma)^2+(na)^2})+$$

$$+\Phi_{M-O}(\sqrt{(x+\ell a+\ell' a)^2+((m+\frac{1}{2})a)^2+((n+\frac{1}{2})a)^2})$$

$$+\Phi_{M-O}(\sqrt{(x+\ell a+(\ell'+\frac{1}{2})a)^2+(ma)^2+((n+\frac{1}{2})a)^2})$$

$$+\Phi_{M-O}(\sqrt{(x+\ell a+(\ell'+\frac{1}{2})a)^2+((m+\frac{1}{2})a)^2+(na)^2})$$

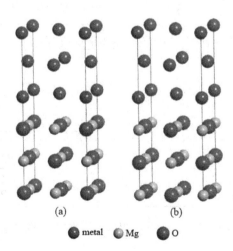

Fig. 6.10 Two ideal structures of fcc(100)/MgO(100)

$$+ \Phi_{M-O}(\sqrt{(x+(\ell+\frac{1}{2})a+\ell'a)^2+(ma)^2+((n+\frac{1}{2})a)^2})$$

$$+ \Phi_{M-O}(\sqrt{(x+\ell a+(\ell'+\frac{1}{2})a)^2+((m+\frac{1}{2})a)^2+(na)^2})$$

$$+ \Phi_{M-O}(\sqrt{(x+(\ell+\frac{1}{2})a+(\ell'+\frac{1}{2})a)^2+(ma)^2+(na)^2})$$

$$+ \Phi_{M-O}(\sqrt{(x+(\ell+\frac{1}{2})a+(\ell'+\frac{1}{2})a)^2+((m+\frac{1}{2})a)^2+((n+\frac{1}{2})a)^2})$$

$$+ \Phi_{M-Mg}(\sqrt{(x+\ell a+\ell'a)^2+(ma)^2+((n+\frac{1}{2})a)^2})$$

$$+ \Phi_{M-Mg}(\sqrt{(x+\ell a+\ell'a)^2+((m+\frac{1}{2})a)^2+(na)^2})$$

$$+ \Phi_{M-Mg}(\sqrt{(x+\ell a+(\ell'+\frac{1}{2})a)^2+(ma)^2+(na)^2})$$

$$+ \Phi_{M-Mg}(\sqrt{(x+\ell a+(\ell'+\frac{1}{2})a)^2+((m+\frac{1}{2})a)^2+((n+\frac{1}{2})a)^2})$$

$$+ \Phi_{M-Mg}(\sqrt{(x+(\ell+\frac{1}{2})a+\ell'a)^2+(ma)^2+(na)^2})+$$

$$+ \Phi_{M-Mg}(\sqrt{(x+(\ell+\frac{1}{2})a+\ell'a)^2+((m+\frac{1}{2})a)^2+((n+\frac{1}{2})a)^2})$$

$$+ \Phi_{M-Mg}(\sqrt{(x+(\ell+\frac{1}{2})a+(\ell'+\frac{1}{2})a)^2+(ma)^2+((n+\frac{1}{2})a)^2})$$

$$+ \Phi_{M-Mg}(\sqrt{(x+(\ell+\frac{1}{2})a+(\ell'+\frac{1}{2})a)^2+((m+\frac{1}{2})a)^2+(na)^2})\}$$

(6.95)

and

$$E_{Mg} = \sum_{\ell,\ell'=0}^{\infty} \sum_{m,n=-\infty}^{\infty} \{\Phi_{M-Mg}(\sqrt{(x+\ell a+\ell'a)^2+(ma)^2+(na)^2})$$

$$+ \Phi_{M-Mg}(\sqrt{(x+\ell a+\ell'a)^2+((m+\frac{1}{2})a)^2+((n+\frac{1}{2})a)^2})$$

$$+ \Phi_{M-Mg}(\sqrt{(x+\ell a+(\ell'+\frac{1}{2})a)^2+(ma)^2+((n+\frac{1}{2})a)^2})$$

$$+ \Phi_{M-Mg}(\sqrt{(x+\ell a+(\ell'+\frac{1}{2})a)^2+((m+\frac{1}{2})a)^2+(na)^2})$$

$$+ \Phi_{M-Mg}(\sqrt{(x+(\ell+\tfrac{1}{2})a+\ell' a)^2 + (ma)^2 + ((n+\tfrac{1}{2})a)^2}$$

$$+ \Phi_{M-Mg}(\sqrt{(x+\ell a+(\ell'+\tfrac{1}{2})a)^2 + ((m+\tfrac{1}{2})a)^2 + (na)^2}$$

$$+ \Phi_{M-Mg}(\sqrt{(x+(\ell+\tfrac{1}{2})a+(\ell'+\tfrac{1}{2})a)^2 + (ma)^2 + (na)^2}$$

$$+ \Phi_{M-Mg}(\sqrt{(x+(\ell+\tfrac{1}{2})a+(\ell'+\tfrac{1}{2})a)^2 + ((m+\tfrac{1}{2})a)^2 + ((n+\tfrac{1}{2})a)^2}$$

$$+ \Phi_{M-O}(\sqrt{(x+\ell a+\ell' a)^2 + (ma)^2 + ((n+\tfrac{1}{2})a)^2}$$

$$+ \Phi_{M-O}(\sqrt{(x+\ell a+\ell' a)^2 + ((m+\tfrac{1}{2})a)^2 + (na)^2}$$

$$+ \Phi_{M-O}(\sqrt{(x+\ell a+(\ell'+\tfrac{1}{2})a)^2 + (ma)^2 + (na)^2}$$

$$+ \Phi_{M-O}(\sqrt{(x+\ell a+(\ell'+\tfrac{1}{2})a)^2 + ((m+\tfrac{1}{2})a)^2 + ((n+\tfrac{1}{2})a)^2}$$

$$+ \Phi_{M-O}(\sqrt{(x+(\ell+\tfrac{1}{2})a+\ell' a)^2 + (ma)^2 + (na)^2}+$$

$$+ \Phi_{M-O}(\sqrt{(x+(\ell+\tfrac{1}{2})a+\ell' a)^2 + ((m+\tfrac{1}{2})a)^2 + ((n+\tfrac{1}{2})a)^2}$$

$$+ \Phi_{M-O}(\sqrt{(x+(\ell+\tfrac{1}{2})a+(\ell'+\tfrac{1}{2})a)^2 + (ma)^2 + ((n+\tfrac{1}{2})a)^2}$$

$$+ \Phi_{M-O}(\sqrt{(x+(\ell+\tfrac{1}{2})a+(\ell'+\tfrac{1}{2})a)^2 + ((m+\tfrac{1}{2})a)^2 + (na)^2})\} \quad (6.96)$$

For convenience, let us introduce

$$E_{\pm}(x) = E_{M-O}(x) \pm E_{M-Mg}(x) \tag{6.97}$$

and

$$\Phi_{\pm}(x) = \Phi_{M-O}(x) \pm \Phi_{M-Mg}(x) \tag{6.98}$$

then Eqs. (6.95) and (6.96) become

$$E_{\pm} = \sum_{\ell,\ell'=0}^{\infty} \sum_{m,n=-\infty}^{\infty} \{\Phi_{\pm}(\sqrt{(x+\ell a+\ell' a)^2 + (ma)^2 + (na)^2})$$

$$+ \Phi_{\pm}(\sqrt{(x+\ell a+\ell' a)^2 + ((m+\tfrac{1}{2})a)^2 + ((n+\tfrac{1}{2})a)^2})$$

$$+ \Phi_\pm(\sqrt{(x+\ell a+(\ell'+\frac{1}{2})a)^2+(ma)^2+((n+\frac{1}{2})a)^2})$$

$$+ \Phi_\pm(\sqrt{(x+\ell a+(\ell'+\frac{1}{2})a)^2+((m+\frac{1}{2})a)^2+(na)^2})$$

$$+ \Phi_\pm(\sqrt{(x+(\ell+\frac{1}{2})a+\ell'a)^2+(ma)^2+((n+\frac{1}{2})a)^2})$$

$$+ \Phi_\pm(\sqrt{(x+\ell a+(\ell'+\frac{1}{2})a)^2+((m+\frac{1}{2})a)^2+(na)^2})$$

$$+ \Phi_\pm(\sqrt{(x+(\ell+\frac{1}{2})a+(\ell'+\frac{1}{2})a)^2+(ma)^2+(na)^2})$$

$$+ \Phi_\pm(\sqrt{(x+(\ell+\frac{1}{2})a+(\ell'+\frac{1}{2})a)^2+((m+\frac{1}{2})a)^2+((n+\frac{1}{2})a)^2})$$

$$\pm \Phi_\pm(\sqrt{(x+\ell a+\ell'a)^2+(ma)^2+((n+\frac{1}{2})a)^2})$$

$$\pm \Phi_\pm(\sqrt{(x+\ell a+\ell'a)^2+((m+\frac{1}{2})a)^2+(na)^2})$$

$$\pm \Phi_\pm(\sqrt{(x+\ell a+(\ell'+\frac{1}{2})a)^2+(ma)^2+(na)^2})$$

$$\pm \Phi_\pm(\sqrt{(x+\ell a+(\ell'+\frac{1}{2})a)^2+((m+\frac{1}{2})a)^2+((n+\frac{1}{2})a)^2})$$

$$\pm \Phi_\pm(\sqrt{(x+(\ell+\frac{1}{2})a+\ell'a)^2+(ma)^2+(na)^2})$$

$$\pm \Phi_\pm(\sqrt{(x+(\ell+\frac{1}{2})a+\ell'a)^2+((m+\frac{1}{2})a)^2+((n+\frac{1}{2})a)^2})$$

$$\pm \Phi_\pm(\sqrt{(x+(\ell+\frac{1}{2})a+(\ell'+\frac{1}{2})a)^2+(ma)^2+((n+\frac{1}{2})a)^2})$$

$$\pm \Phi_\pm(\sqrt{(x+(\ell+\frac{1}{2})a+(\ell'+\frac{1}{2})a)^2+((m+\frac{1}{2})a)^2+(na)^2})\} \quad (6.99)$$

The above equations can be rewritten as

$$E_\pm = \sum_{\ell,\ell'=0}^{\infty} \sum_{m,n=-\infty}^{\infty} \Phi_\pm(\sqrt{(x+\frac{(\ell+\ell')a}{2})^2+\frac{(m^2+n^2)a^2}{4}}) \quad (6.100)$$

In order to solve $\Phi_\pm(x)$ based on ab initio $E_\pm(x)$, let us define a function $H_\pm$ as

$$H_\pm(x) = \sum_{m,n=-\infty}^{\infty} \Phi_\pm(\sqrt{x^2+\frac{(m^2+n^2)a^2}{4}}) \quad (6.101)$$

Table 6.16 The coefficients h(k) and g(k) for a Metal(bcc)/MgO(001) interface

k	h(k)	g(k)
1	1	1
2	4	−4
3	4	12
4	0	−32
5	4	76
6	8	−168
7	0	352
8	0	−704
9	4	1356
10	4	−2532
11	8	4600
12	0	−8160

Then Eq. (6.100) becomes

$$E_\pm(x) = \sum_{\ell,\ell'=0}^{\infty} (\pm 1)^{\ell+\ell'} H_\pm(x + (\ell+\ell')a/2) \qquad (6.102)$$

Equation (6.102) includes double summations which correspond to double Möbius inversions, then

$$H_\pm(x) = E_\pm(x) \mp 2E_\pm(x + \frac{a}{2}) + E_\pm(x + a) \qquad (6.103)$$

Equation (6.101) can be also written as

$$H_\pm(x) = \sum_{n=0}^{\infty} (\pm 1)^k h(k) \Phi_\pm(\sqrt{x^2 + n\frac{a^2}{4}}) \qquad (6.104)$$

By using additive Cesàro inversion

$$\Phi_\pm(x) = \sum_{n=0}^{\infty} (\pm 1)^n g(n) H_\pm(\sqrt{x^2 + n\frac{a^2}{4}}) \qquad (6.105)$$

where $g(n)$ can be determined by recursion

$$\sum_{n=0}^{k} h(k-n) g(n) = \delta(k,0) \qquad (6.106)$$

Go back to the coefficient $h(k)$, from (6.101) to (6.104), the double summation with m, n over $\pm \infty$ becomes a single summation from 0 to ∞, the coefficient $h(k)$ represents the number of integer solution of equation $k = m^2 + n^2$. This is just the same as coordination number in 2D square lattice. Note that the number of solutions is equal to 1 when $k = 0$. Thus for a metal with fcc structure in M/MgO interface, we have

$$\Phi_\pm(x) = \sum_{n=0}^{\infty} g(n)\{E_\pm(\sqrt{x^2 + \frac{4a^2}{4}})$$

$$\mp 2E_\pm(\sqrt{x^2 + \frac{4a^2}{4}} + a) + E_\pm(\sqrt{x^2 + \frac{4a^2}{4}} + 2a) \quad (6.107)$$

Note that in order to ensure the convergence of the sums in Eqs. (6.100) and (6.107), the Rahman-Stillinger-Lamberg function

$$\Phi(r) = D_0 e^{y(1-\frac{r}{R_0})} + \frac{a_1}{1 + e^{b_1(r-c_1)}} + \frac{a_2}{1 + e^{b_2(r-c_2)}} + \frac{a_3}{1 + e^{b_3(r-c_3)}} \quad (6.108)$$

is chosen for fitting the calculated data series of both $E(x)$ and $\Phi(x)$. Some examples are shown in Table 6.17.

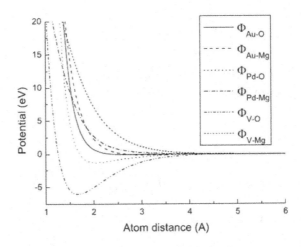

Fig. 6.11 Several interatomic potentials across Metal/MgO(001) interface.

For a *bcc* metal in M/MgO, the formula becomes

$$\Phi_\pm(x) = \sum_{n=0}^{\infty} g(n)\{E_\pm(\sqrt{x^2 + \frac{4a^2}{4}}) \mp E_\pm(\sqrt{x^2 + \frac{4a^2}{4}} + a)$$

$$\mp E_\pm(\sqrt{x^2 + \frac{4a^2}{4}} + \frac{a}{\sqrt{2}}) + E_\pm(\sqrt{x^2 + \frac{4a^2}{4}} + 2a) \quad (6.109)$$

Table 6.17 RSL parameters of interatomic potentials across M/MgO(100) interfaces (M=Ag, Al, Au, Cu)

	D_0 eV	R_0 A	y	a_1 eV	b_1 A^{-1}	c_1 A	a_2 eV	b_2 A^{-1}	c_2 A	a_3 eV	b_3 A^{-1}	c_3 A
Φ_{Ag-O}	242.83	1.00	2.14	−246.39	2.84	1.08	−27.55	2.11	1.90	−0.32	2.58	3.55
Φ_{Ag-Mg}	12.35	1.00	2.08	69.47	6.24	1.15	0.31	3.66	2.98	3.09	4.38	1.90
Φ_{Al-O}	43.52	1.00	3.26	−83.01	2.46	0.70	−0.40	2.26	3.07	1.96	1.51	1.09
Φ_{Al-Mg}	60.49	1.00	1.08	−5.43	3.53	1.92	−11.19	1.90	2.26	−11.48	1.08	2.53
Φ_{Au-O}	290.30	1.00	2.61	−157.34	3.78	1.16	−89.78	2.31	1.31	0.10	2.17	3.93
Φ_{Au-Mg}	170.36	1.00	2.36	−753.00	2.21	0.07	−6.53	4.46	1.71	−0.88	3.66	2.46
Φ_{Cu-O}	52.98	1.00	2.05	−26.60	3.49	1.40	11.85	2.70	2.30	−22.09	2.44	2.19
Φ_{Cu-Mg}	6.23	1.00	4.04	13.04	4.43	1.42	2.90	2.95	2.07	0.14	3.39	3.24

There are several interatomic potentials between atoms across the interface M/MgO shown in Fig. 6.11. When the orientation of intersecting surface in the metal part is (111), one has to change a few of atomic coordinations a little bit to form a semigroup. The similar situations happen for M/Al$_2$O$_3$(111)[Lon2007], Al(001)/3C-SiC(001)[Zha2008], Al/SiC(111)[Zha2009].

Now let us introduce an interpolation method to avoid both series convergence and structure complicity problems. As shown in Fig. 6.12, there are two models: one is with Mg terminated for MgO(111) surface, other is with O terminated. Therefore, the adhesive energies $E_O(x)$ and $E_{Mg}(x)$ can be expressed as

$$E_O(x) = \sum_{i,j} \Phi_{Ag-O}(r_{i,j}^{Ag-O}(x)) + \sum_{i,j} \Phi_{Ag-Mg}(r_{i,j}^{Ag-Mg}(x)) \quad (6.110)$$

$$E_{Mg}(x) = \sum_{i,j} \Phi_{Ag-O}(r_{i',j'}^{Ag-O}(x)) + \sum_{i,j} \Phi_{Ag-Mg}(r_{i',j'}^{Ag-Mg}(x)) \quad (6.111)$$

where x its the interfacial distance, i, i' represents the atomic position in MgO part, j, j' represents the atomic position in Ag. $r_{i,j}^{Ag-O}(x), r_{i,j}^{Ag-Mg}(x), r_{i',j'}^{Ag-O}(x), r_{i',j'}^{Ag-Mg}(x)$ represent the distances between Ag atoms and ions, and these distances are also dependent on interfacial distance x.

For convenience, let separate the range of interfacial distances x into N equal parts such that

$$x_{min} = x_0 < x_1 < x_2 < ... < x_N = x_{max} \quad (6.112)$$

Noted that N is a big number. Simultaneously the range of distances $r_{i,j}^{Ag-O}(x), r_{i,j}^{Ag-Mg}(x), r_{i',j'}^{Ag-O}(x), r_{i',j'}^{Ag-Mg}(x)$ is also separated into N parts

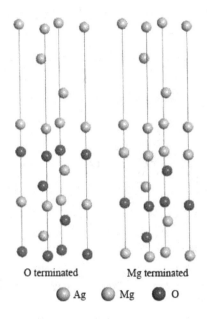

Fig. 6.12 Polar interface Ag(111)/MgO(111)

as
$$r_{min} = r_0 < r_1 < r_2 < \ldots < r_N = r_{max} \tag{6.113}$$

For an arbitrary distance x, there exists a linear interpolation for $\Phi_{Ag-O}(r_{i,j}^{Ag-O}(x))$ such that

$$\begin{aligned}\Phi_{Ag-O}(r_{i,j}^{Ag-O}(x)) &\approx \frac{r_{m+1} - r_{i,j}^{Ag-O}(x)}{r_{m+1} - r_m}\Phi_{Ag-O}(r_m) \\ &+ \frac{r_{i,j}^{Ag-O}(x) - r_m}{r_{m+1} - r_m}\Phi_{Ag-O}(r_{m+1}) \\ &= a_m \Phi_{Ag-O}(r_m) + a_{m+1}\Phi_{Ag-O}(r_{m+1})\end{aligned} \tag{6.114}$$

in which $r_{m+1} \geq r_{i,j}^{Ag-O}(x) \geq r_m$. This approximation will be accurate enough when $|r_{m+1} - r_m|$ is very small. Similarly, we have

$$\begin{aligned}\Phi_{Ag-Mg}(r_{i,j}^{Ag-Mg}(x)) &\approx \frac{r_{m+1} - r_{i,j}^{Ag-Mg}(x)}{r_{m+1} - r_m}\Phi_{Ag-Mg}(r_m) \\ &+ \frac{r_{i,j}^{Ag-Mg}(x) - r_m}{r_{m+1} - r_m}\Phi_{Ag-Mg}(r_{m+1}) \\ &= a_m \Phi_{Ag-Mg}(r_m) + a_{m+1}\Phi_{Ag-Mg}(r_{m+1})\end{aligned} \tag{6.115}$$

Therefore, Eqs. (6.110) and (6.111) become

$$E_O(x_n) = \sum_m A_{n,m}\Phi_{Ag-O}(r_m) + \sum_m B_{n,m}\Phi_{Ag-Mg}(r_m) \quad (6.116)$$

$$E_{Mg}(x_n) = \sum_m C_{n,m}\Phi_{Ag-O}(r_m) + \sum_m D_{n,m}\Phi_{Ag-Mg}(r_m) \quad (6.117)$$

Now we define

$$\vec{E}_O = [E_O(x_1), E_O(x_2), ..., E_O(x_N)] \quad (6.118)$$

$$\vec{E}_{Mg} = [E_{Mg}(x_1), E_{Mg}(x_2), ..., E_{Mg}(x_N)] \quad (6.119)$$

$$\vec{\Phi}_{Ag-O} = [\Phi_{Ag-O}(x_1), \Phi_{Ag-O}(x_2), ..., \Phi_{Ag-O}(x_N)] \quad (6.120)$$

$$\vec{\Phi}_{Ag-Mg} = [\Phi_{Ag-Mg}(x_1), \Phi_{Ag-Mg}(x_2), ..., \Phi_{Ag-Mg}(x_N)] \quad (6.121)$$

where $\vec{E}_O \vec{E}_{Mg} \vec{\Phi}_{Ag-O} \vec{\Phi}_{Ag-Mg}$ are Nth order vectors, A,B,C and D are N×N matrices. Than Eqs. (6.116) and (6.117) can be denoted simply by

$$[\vec{E}_O, \vec{E}_{Mg}] = [\vec{\Phi}_{Ag-O}, \vec{\Phi}_{Ag-Mg}] \begin{bmatrix} A & B \\ C & D \end{bmatrix} \quad (6.122)$$

This is a typical set of linear equations, the non-zero element distribution is shown in Fig. 6.13. The solution of Eq. (6.122) can be obtained by matrix inversion as

$$[\vec{\Phi}_{Ag-O}, \vec{\Phi}_{Ag-Mg}] = [\vec{E}_O, \vec{E}_{Mg}] \begin{bmatrix} A & B \\ C & D \end{bmatrix}^{-1} \quad (6.123)$$

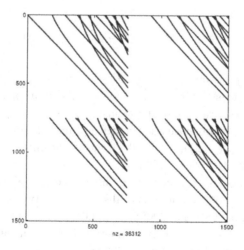

Fig. 6.13 Non-zero element distribution in the above matrix.

Table 6.18 RSL parameters of interatomic potentials across Ag/MgO(100) interfaces

	D_0(eV)	R_0(A)	y	a_1(eV)	$b_1(A^{-1})$	c_1(A)
Φ_{Ag-O}	167.73	1.00	1.57	−64.03	3.15	1.43
Φ_{Ag-Mg}	167.17	1.00	1.58	−63.64	2.94	1.36
	a_2(eV)	$b_2(A^{-1})$	c_2(A)	a_3(eV)	$b_3(A^{-1})$	c_3(A)
Φ_{Ag-O}	−78.87	1.60	1.54	0	0	0
Φ_{Ag-Mg}	−81.24	1.50	1.45	0.06	2.39	5.85

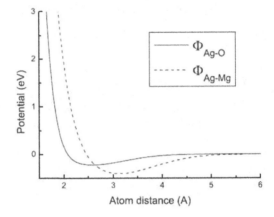

Fig. 6.14 Interatomic potentials across Ag(111)/MgO(111) interface.

This method implies cutoff at first, thus there is no convergence problem. Second, it does not requires a special distribution of atomic positions, thus there is no requirement to construct a semigroup. Third, RSL2-type functions are again chosen as the fitting functions for calculated data of $\vec{E}_O \vec{E}_{Mg} \vec{\Phi}_{Ag-O} \vec{\Phi}_{Ag-Mg}$. Fig. 6.14 shows the interatomic potentials across $Ag(111)/MgO(111)$ interface.

The above interfacial potentials have been applied to the misfit dislocations[Lon2005b] and misfit dislocation networks on some typical interfaces[Lon2008b; Lon2008d; Lon2009], also, to the interfacial Burgers vectors, the morphology of metal clusters on surface of MgO, [Lon2008g; Lon2008f; Lon2008e; Zha2009a], interface fracture[Wei2008], and so on.

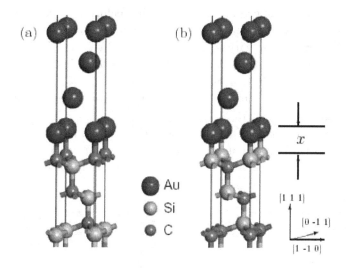

Fig. 6.15 Ideal Au(111)/SiC(111) interface models used for the inversion method. (a) C-terminated, (b) Si-terminated. Each model consists of six metal layers, six Si layers and six C layers. Only a few layers near interface are shown here.

6.10.3 *Matal/SiC interface*

Silicon carbide (SiC) as one of the most promising semiconductors for high temperature, high-frequency and high-power application, is attracting more and more attentions for its wide band gap and excellent performance. In device manufacture, metal/SiC interface is an essential issue for its key role in determining the properties of the devices. For example, the metal-SiC junction can be either Ohmic or Schottky contact for different metals.

Table 6.19 RSL parameters of two-body potentials across M/SiC(111) interfaces (M=Au, Ag, Pt, Al)

	Φ_{Au-C}	Φ_{Au-Si}	Φ_{Ag_C}	Φ_{Ag-Si}	Φ_{Pt-C}	Φ_{Pt-Si}	Φ_{Al-C}	Φ_{Al-Si}
$D_0(eV)$	20.5909	30.0401	20.2023	30.1051	21.2815	30.2722	0.1392	68.4853
$R_0(\text{Å})$	1.000	1.000	1.000	1.000	1.000	1.000	1.000	1.000
y	2.0784	1.3763	2.0358	1.3588	2.6728	1.5170	5.7526	1.2844
$a_1(eV)$	2.4716	−0.0330	1.2585	0.0054	7.7074	−0.2374	66.9374	75.0606
$b_1(\text{Å}^{-1})$	10.9813	9.2316	12.8937	9.6162	8.1985	3.4732	5.7654	2.9238
$c_1 \text{Å}$	1.7318	3.9789	1.7386	4.0720	1.5418	4.0549	0.9401	0.9512
$a_2(eV)$	−6.9938	−14.7021	−6.9290	−14.6264	−7.0506	−14.7741	1.1365	−12.2142
$b_2(\text{Å}^{-1})$	2.0705	1.4725	1.9743	1.4300	2.0590	1.5953	8.2083	3.6486
$c_2(\text{Å})$	1.8643	1.9144	1.8758	1.8660	1.6342	1.8372	1.6578	1.6732
$a_3(eV)$	3.3455	7.0814	3.3737	4.3918	3.3618	11.3124	−0.6407	−55.5837
$b_3(\text{Å}^{-1})$	1.4406	7.9455	1.4554	8.1147	1.5500	6.2888	2.3362	1.3456
$c_3(\text{Å})$	1.0634	1.7157	1.0779	1.7515	0.9233	1.6075	2.7657	1.4086

Considering the large computational cost in dealing with complex atomic models as well as dynamic processes, some researchers have paid attention to interatomic potentials instead [Luo99; Zha2008]. Recently, Zhao and Wang applied the Chen-Möbius inversion method[Zha2009; Wan2010] to $Metal/SiC(111)$ interfaces. First, they obtain two-body potentials form two ideal polar interface models: one is Si-terminated, another is C-terminated(see Fig. 6.15).

The deduction and calculation of two-body potentials are similar to that for Eq. (6.123), the calculated RSL parameters are listed in Table 6.18, in which the potentials for Au/SiC(111) are shown in Fig. 6.16.

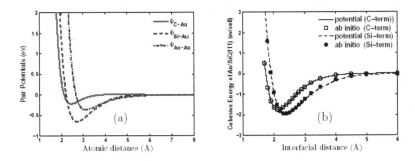

Fig. 6.16 (a)Calculated pair potentials. (b)ab initio adhesion comparing to that from potentials. Here Φ_{Au-Au} for bulk Au is also presented for use in metal slab.

In Fig. 6.17, taking the C-terminated Au/SiC case as an illustration, only a few layers near the interface are presented. (b) metal(111) surface (top view) with Burgers vectors of perfect dislocation b1 and partial dislocations b2, b3. Metal atoms are marked as A, B and C when they are in top site, hollow site and hex site according to the SiC(111) surface. They have calculated the adhesive energy based on interfacial pair potentials, the result is in good agreement with *ab initio* adhesive curve. However, the bonding orientation of atoms on the surface of SiC is very strong, the three-body potential must be required to fit various interface configurations(see Fig. 6.17). For this purpose, The Stillinger–Weber three-body potential (SW3) is introduced as a corrected term, that is

$$\phi_{jik}(r_{ij}, r_{ik}, \theta_{jik}) = \lambda_{jik} exp[\frac{\nu_{ij}}{r_{ij} - R_{ij}} + \frac{\nu_{ik}}{r_{ik} - R_{ik}}]$$
$$\times (\cos\theta_{jik} - \cos\theta_0) \qquad (6.124)$$

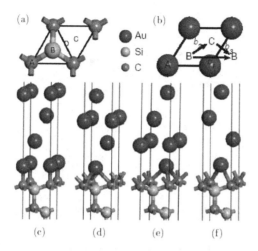

Fig. 6.17 Atomic configurations of the metal(111)/SiC(111) interface. (a) SiC(111) surface (top view). The symbols A, B, C and D refer to top site, hollow site, hex site and bridge site, respectively. (c), (d), (e) and (f) are top-site, hollow-site, hex-site and bridge-site structures (side view), respectively.

Table 6.20 Stillinger–Weber parameters of three-body potentials across M/SiC(111) interfaces (M=Au, Ag, Pt, Al)

$j - i - k$	λ_{jik}(eV)	θ_0°	ν_{ij}(A)	ν_{ik}(A)	R_{ij}(A)	R_{ik}(A)
$Au - Si - C$	0.8000	109.47	0.2300	0.2047	3.7500	3.1832
$Au - Si - Au$	2.2000	51.00	0.2300	0.2300	3.7500	3.7500
$Au - C - Si$	0.4250	109.47	0.2000	0.2047	3.5500	3.1832
$Au - C - Au$	2.8000	45.00	0.2000	0.2000	3.5500	3.5500
$Ag - Si - C$	0.3000	109.47	0.2000	0.2047	3.7500	3.1832
$Ag - Si - Ag$	1.1300	35.00	0.2000	0.2000	3.7500	3.7500
$Ag - C - Si$	0.1000	109.47	0.2700	0.2047	3.5500	3.1832
$Ag_C - Ag$	0.9000	5.00	0.2700	0.2700	3.5500	3.5500
$Pt - Si - C$	1.2500	109.47	0.2000	0.2047	3.5200	3.1832
$Pt - Si - Pt$	0.1500	30.000	0.2000	0.2000	3.5200	3.5200
$Pt - C - Si$	0.8500	109.47	0.2000	0.2047	3.4600	3.1832
$Pt - C - Pt$	0.6750	6.25	0.2000	0.2000	3.4600	3.4600
$Al - Si - C$	0.9109	109.47	0.1878	0.2047	3.6500	3.1832
$Ai - Si - Al$	0.0342	180.00	0.1878	0.1878	3.6500	3.6500
$Al - C - Si$	0.3283	109.47	0.0584	0.2047	3.4300	3.1832
$Al - C - Al$	0.6229	2.30	0.0584	0.0584	3.4300	3.4300

Based on both pair- and 3-body potentials cross interface with the the interatomic potentials in metal slab and in SiC slab, the interfacial structure of various configurations can be described. For example, the dislocation core is positioned in the first metal layer (P=1) for M=(Ag, Pt), and

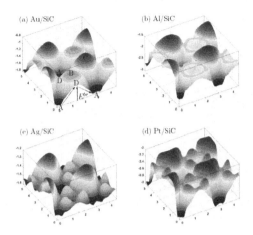

Fig. 6.18 C-terminated AE distribution of metal(111)/SiC(111) interfaces(metal=Au, Al, Ag and Pt).

appears in metal slabs (P>1) for M=(Au, Al). This will affect the interfacial fracture properties significantly. There are also many differences between metal/SiC(001), metal/SiC(001) and metal/SiC(111). Also, interatomic potentials can be used to discuss misfit dislocation network and dislocation decomposition. Note that when metal slab slips on SiC substrate, there are many interfacial configurations, the distribution map for the adhesive energies would be interesting. At the end of this section, the distribution map for adhesive energy for various interfacial configurations for different metals is shown in Fig. 6.18[Wan2010]. Parallel to interface, the coordinate changing in x-y plane expresses different interfacial configuration. The z-axis represents adhesive energy (units: eV/unitcell). Symbols A, B, C, D denote sites when metal slab is positioned at top site, hollow site, hex site and bridge site, respectively. In conclusion, this section proves a new interface inversion by demonstration, with a string of example only *ab initio* solved with difficulty.

6.11 Summary

Chapter 6 mainly presents the Möbius inversion formula for a general 3D lattice in order to extract interatomic potentials between like and unlike atoms in solids or clusters. This chapter includes the following points.

(1) The interatomic distance function set $\{b(n)\}$ without closure re-

garding common multiplication operations was extended to a multiplicative semigroup $\{B(n)\}$, then a generalized Möbius inversion formula was established. This can be applied to extracting pairwise interatomic potentials from *abinitio* cohesive energy curves. The Chapter 6 provides many concrete inversion expressions for a variety of lattice structures such as fcc, bcc, $L1_2$, DO_3, B_1, B_2 and B_3 structures. The lattice inversion method has been applied to different material systems with distinct bondings such as the intermetallic compounds, the ionic compounds, the III-V semiconductor compounds, the metallic carbides and nitrites, rare-earth intermetallic compounds.

(2)A modified lattice inversion formula can be stated as follows.

$$F(n,x) = \sum_{B(d)|B(n)} r(d) f[B^{-1}\frac{B(n)}{B(d)}, B(d)x]$$

$$\Leftrightarrow f(n,x) = \sum_{B(d)|B(n)} I(d) F[B^{-1}\frac{B(n)}{B(d)}, B(d)x] \qquad (6.125)$$

Where one of new formulas

$$F(n) = \sum_{B(d)|B(n)} r(d) f[B^{-1}\frac{B(n)}{B(d)}]$$

$$\Leftrightarrow f(n) = \sum_{B(d)|B(n)} I(d) F[B^{-1}\frac{B(n)}{B(d)}] \qquad (6.126)$$

might be useful for further study.

(3)The application of lattice inversion method to improving the embedded atom method was presented, this may simplify the complicated procedure significantly for the conventional EAM.

(4)The Möbius–Cesàro inversion formula has been applied to extracting atom–atom, ion–ion and ion–atom interactions across interfaces based on an additive semigroup analysis with a number of applications to metal/MgO, metal/Al2O3 and other interfacial systems. Note that a more general interfacial structures may transcend the semigroup case.

(5)The cluster expansion method (CEM) for extracting many-body interactions in clusters and solids was introduced in detail by using the Möbius inverse formula in a partially ordered set. The effective combination of the general lattice inversion formula for pair potential and the cluster expansion method is an additional subject for future study.

(6)The convergence problem of a infinite series is replaced by selecting of fitting function of interatomic potential. The most important thing we have to remember is as Abel's saying that "Until now the theory of infinite series in general has been very badly grounded. One applies all the operations to infinite series as if they were finite; but is that permissible? I think not. Where is it demonstrated that one obtains the differential of an infinite series by taking the differential of each term? Nothing is easier than to give instances where this is not so. "[Rem91].

Note that for dealing with a periodic structure, both congruence analysis and Fourier transform are convenient and efficient. From the hidden multiplicative semigroup to the invisible additive semigroup, we always focus on the algebraic structure of the physical system under consideration. Fortunately, many inverse lattice inversion problems or their core parts in the problems can be solved by applying these methods without proving the convergence.

The history of constructing interatomic potentials can be traced back to the early attempt of the 1930s. In the many years since then, great progress has been made in both theory and application, from various pair potential models to multi-body ones, from simple metals to ionic crystals with charge transfer, from ideal crystals to defective systems. Nowadays, not surprisingly, as computer-processing power increases exponentially and progress is made in fully quantum-mechanical treated methods, the effort to achieve more accurate and more practicable atomic potential models is not diminishing, rather, it is becoming more intense than ever before. This is due in part, also, to a high demand for people to pursue the larger scale and multi-scale atomistic simulations of complicated materials behaviors. Many activities in extending present models and developing new approaches are underway, bringing new vitality to this area and at the same time various difficulties[Che2008]. For example, EAM has been in existence for 25 years, but the first and second nearest neighbor modified versions are still in discussion [Lee2000; Lee2001; Lee2003; Kim2006; Bas2007]. Maybe the lattice inversion with semigroup analysis will reduce some of these shortcomings in conventional EAM.

It is concluded that the lattice inversion method is useful to obtain interatomic potentials systematically, in particular, for the complex systems, which are otherwise difficult to deal with.

Appendix: Möbius Inverse Formula on a Partially Ordered Set

In the previous chapters we introduce a variety of Möbius inverse formulas on a multiplicative semigroup or on an additive semigroup. Möbius inverse formula on a partially ordered set can be considered an extension of the previous Möbius inverse formulas. No matter which branch in mathematics we are interested in, we have to determine the objects of our research; these objects form a set, and between the elements in the set there are some special relations or operations. This section provides some elementary knowledge on partially ordered sets and Möbius inversion on a partially ordered set. The partially ordered set will be very useful for the reader's further study[Rot64; Lin92].

A.1 TOSET

First of all, let us introduce the definition of a totally ordered set.

Definition A.1. Assuming that S is a nonempty set with a binary relation $\leq$ (*sometimes* $\subseteq$). If
 (i) $a \leq a$ for all $a \in S$ (*reflexivity*),
 (ii) if $a \leq b$ and $b \leq c$, then $a \leq c$ (*transitivity*),
 (iii) if $a \leq b$ and $b \leq a$, then $a = b$ (*antisymmetry*),
 (vi) for any $a, b \in S$, either $a \leq b$ or $b \leq a$,
then S is called a totally ordered set (TOSET)[1].

---
[1] Note that if $a \leq b$ and $a \leq b$, it is denoted as $a < b$. If there is a binary relation between a and b, and a does not satisfy $a \leq b$, then it is denoted as $a \nleq b$, which means either there is no binary relation between a and b, or $a > b$.

Fig. A.1 Gian Carlo Rota (1932–1999) in 1970.

A.2 POSET

If a binary relation $\leq$ is only available for part of elements in a nonempty set S, then we consider a partially ordered set.

Definition A.2. Let P is a nonempty set with a binary relation $\leq$ (*sometimes* $\subseteq$) such that
 (i) $a \leq a$ for all $a \in S$ (*reflexivity*),
 (ii) if $a \leq b$ and $b \leq c$, then $a \leq c$ (*transitivity*),
 (iii) if $a \leq b$ and $b \leq a$, then $a = b$ (*antisymmetry*),
then P is called a partially ordered set(POSET).

Example 1. Define a set $S_1 = \{1, 2, 3, 4, 6, 12\}$ with a binary relation $a \leq b$ when $a|b$, then we have

$$1 \leq 2 \leq 4 \leq 12, \text{ and } 1 \leq 3 \leq 6 \leq 12, \text{ and } 2 \leq 6$$

Note that there is no binary relation between 2 and 3, between 3 and 4, between 4 and 6. In other words, according to above definition on the binary relation, these pairs are incomparable. S_1 is a POSET[2].

Example 2. Define a set $S_2 = \{1, 2, 3, 4, 6, 12\}$ and all its subsets, with binary relation $a \leq b$ when $a \subset b$. Therefore, we have

$$\{1,2\} \subset \{1,2,3\} \subset \{1,2,3,4\} \text{ and } \{1,2\} \subset \{1,2,4\} \subset \{1,2,3,4\}$$

Note that there is no binary relation between the following subsets such as $\{1,2\}, \{1,3\}, \{1,4\}, \{2,6\}, \{3,4,12\}$. Therefore, S_2 is a POSET.

Example 3. If we define the set $S = 1, 2, 3, 4, 6, 12$ with binary relation $a \leq b$ when $a < b$ in common sense, then we have

$$1 \leq 2 \leq 3 \leq 4 \leq 6 \leq 12$$

There is a binary relation between any two elements in S, thus S is a TOSET. Note that the property of a set is not only dependent on its composition but also dependent on what kind of binary operation is defined.

Example 4. Let S be a set, and $\mathbb{P}(S)$ is the *powerset* of S, i.e. $\mathbb{P}(S)$ consists of all subsets of S. Now we define $A \leq B$ if $A, B \subset \mathbb{P}(S)$ and $A \subseteq B$. Thus $\mathbb{P}(S)$ becomes a POSET. If S is a finite one, then S and empty set $\emptyset$ are the maximum element and minimum element in $\mathbb{P}(S)$ respectively.

When S is an infinite set, let $\mathbb{P}_f(S)$ is a set including all finite subsets of S. Now we define $A \leq B$ if $A \subseteq B$ and $A, B \subset \mathbb{P}_f(S)$, then $\mathbb{P}_f(S)$ is also a POSET. In this case, there is no maximum in $\mathbb{P}(S)$, but there is still $\emptyset$ as the minimum.

When $x \leq y$ and $x \neq y$, it is denoted as $x < y$.

A.3 Interval and Chain

(1) For many applications, it is convenient to introduce *interval* as follows. For $x, y \in P$ and $x \leq y$, the *interval* is over only those z such that $[x, y] := \{x \leq z \leq y\}$.

In example 1, let $x = 1$ and $y = 6$, the the interval between x and y is

$$[x, y] := \{1, 2, 3, 6\}$$

In Example 2, let $x = \{1, 2\}$ and $y = \{1, 2, 3, 4\}$, then

$$[x, y] = \{\{1, 2\}, \{1, 2, 3\}, \{1, 2, 4\}, \{1, 2, 3, 4\}\}$$

(2) Assuming that $x < y$ and $x, y \in P$. If

$$x = x_0 < x_1 < ... < x_i < x_{i+1} < ... < x_{n-1} < x_n = y \qquad (A.1)$$

---
[2] Correspondingly, the set of operators $\mathcal{S}_1^* = \{T_1, T_2, T_3, T_4, T_6, T_{12}\}$ with $T_n f(x) = f(nx)$ is also a POSET, if we define $T_m \leq T_n$ when $m|n$.

then the set $\{x = x_0, x_1, ..., x_i, x_{i+1}, ..., x_{n-1}, x_n = y\}$ is called a *chain*. Also, if for any nearest neighbor pair (x_i, x_{i+1}) in a chain there is no $z \in P$ such that

$$x_i < z < x_{i+1} \qquad (A.2)$$

then this chain is a close-packed chain, called a *maximumchain* between x and y. It is worth to mention that if there is $z \in P$ such that $x_i < z < x_{i+1}$ for any $i + 1 \leq n$, then this chain can be refined. Obviously, a maximum chain can not be refined any more. The length of a chain between x and y, d(x,y), is defined as the length of any one *maximum chain* between x and y.

In Example 1, let $x = 1, y = 6$, then $d(1,6) = 3$. Note that between x and y there are two maximum chains $\{1, 2, 6\}$ and $\{1, 3, 6\}$.

Note that let us set $T_{b(n)}f(x) = f(b(n)x)$, if for arbitrary m and n, there is always k such that $b(m)b(n) = b(k)$, then the set $S_1 = \{b(n)\}$ and $S_2 = \{T_{b(n)}\}$ are POSET with definition of $b(m) \leq b(n)$ when $b(m)|b(n)$. There is another more common case, that is, for any m and n, there is not always a k such that $b(m)b(n) = b(k)$, then the set $S = \{T_{b(n)}\}$ is not a POSET. In this case, we can refine this set as follows. Let us work with the multiplications of all possible products of elements of S, and construct a new set $S' = \{B(n)\}$ containing all the elements of S and all these products of them. Now we can define $B(m) \leq B(n)$ when $B(m)|B(n)$ in this multiplicative semigroup $\{B(n)\}$. Now the set S has been refined or enlarged as S' and S' is a POSET.

A.4 Local Finite POSET

Definition A.3. Assuming that P is a POSET. If any interval $[x, y]$ with $x, y \in P$ is a finite set, then P is called a locally finite POSET.

If P is a finite set, then P is a finite POSET. Obviously, a finite POSET must be a locally finite POSET. However, a locally finite POSET may be an infinite set. Note that If P is a locally finite POSET, for $x, y \in P$ and $x < y$, then $d(x, y) < \infty$.

Example 4. Assuming S is a set, and $\mathbb{P}(S)$ is the **power set** of S, i.e. $\mathbb{P}(S)$ consists of all subsets of S. Now we define $A \geq B$ if $A, B \subset \mathbb{P}(S)$ and $A \supseteq B$. Thus $\mathbb{P}(S)$ becomes a POSET. If S is a finite one, then S and empty set $\emptyset$ are the maximum element and minimum element in $\mathbb{P}(S)$ respectively.

When S is an infinite set, let $\mathbb{P}_f(S)$ is a set including all finite subsets of S. Now we define $A \geq B$ if $A \supseteq B$ and $A, B \subset \mathbb{P}_f(S)$, then $\mathbb{P}_f(S)$ is also a POSET. In this case, there is no maximum in $\mathbb{P}(S)$, but there is still $\emptyset$ as the minimum.

A.5 Möbius Function on Locally Finite POSET

Definition A.4. Assuming P is a locally finite POSET, R is a commutative ring with unit element, and $\mu(x,y)$ is a binary function defined on P and taking value from R. If $\mu(x,y)$ satisfies the following conditions:

(i) $\mu(x,x) = 1$ for any $x \leq P$

(ii) $\mu(x,y) = 0$ for $x \not\leq y$

(iii) $\sum_{x \leq z \leq y} \mu(x,z) = 0$

then $\mu(x,y)$ is called Möbius function on P.

From (i), (ii) and (iii), it can be proven that

(iv) $\sum_{x \leq z \leq y} \mu(z,y) = 0$

Conversely, (iv) can be proven by (i), (ii) and (iv). The proof can be found[3].

For constructing the Möbius function $\mu(x,y)$, the conditions (i), (ii), (iii) and (iv) have to be used. When $d(x,y) = 0$, it corresponds to $\mu(x,x) = 1$. When $d(x,y) > 0$, for $x, y \in P$ and $x \leq z < y$, then $d(x,z) < d(x,y)$, i.e., $\mu(x,z)$ has been defined. Thus $\mu(x,y)$ is determined as

$$\mu(x,y) = - \sum_{x \leq z < y} \mu(x,z) \tag{A.3}$$

---

[3] Let P is a locally finite POSET, and $x, y \in P$ and $x \leq y$, then interval $[x,y]$ is finite, and denote $|[x,y]| = n$ This is corresponds to $d(x,y) = n-1$. Thus the elements in $[x,y]$ can be put in order as

$$x = x_1, x_2, ..., x_n = y$$

such that $x_i < x_j$ implies $i < j$.

Now we define two matrices A and B with corresponding elements

$$a_{ij} = \mu(x_i, x_j) \text{ if } 1 \leq i, j \leq n$$

$$b_{ij} = \begin{cases} 1 & \text{if } x_i \leq j \\ 0 & \text{if } x_i \not\leq j \end{cases}$$

That is, both A and B are $n \times n$ matrix on R.

From (i), (ii) and (iii), we have $AB = I$, where I is the $n \times n$ unit matrix. From matrix theorem, we have $BA = I$ too, which corresponds to (i), (ii) and (iv).

By using this formula, the Möbius function $\mu(x,y)$ can be evaluated by recursion[4].

A.5.1 *Example*

Let us consider $P = \{1,2,3,4,6,12\}$ and define $x \leq y$ if $x|y$. Now let us construct

$$\mu(x,y) \equiv \begin{pmatrix} \mu(1,1) & \mu(1,2) & \mu(1,3) & \mu(1,4) & \mu(1,6) & \mu(1,12) \\ \mu(2,1) & \mu(2,2) & \mu(2,3) & \mu(2,4) & \mu(2,6) & \mu(2,12) \\ \mu(3,1) & \mu(3,2) & \mu(3,3) & \mu(3,4) & \mu(3,6) & \mu(3,12) \\ \mu(4,1) & \mu(4,2) & \mu(4,3) & \mu(4,4) & \mu(4,6) & \mu(4,12) \\ \mu(6,1) & \mu(6,2) & \mu(6,3) & \mu(6,4) & \mu(6,6) & \mu(6,12) \\ \mu(12,1) & \mu(12,2) & \mu(12,3) & \mu(12,4) & \mu(12,6) & \mu(12,12) \end{pmatrix} \quad (A.5)$$

According to (i), it is given that

$$\mu(1,1) = \mu(2,2) = \mu(3,3) = \mu(4,4) = \mu(6,6) = \mu(12,12) = 1$$

For example, we have

$$\mu(1,2) = -\mu(1,1) = -1$$
$$\mu(1,3) = -\mu(1,1) = 1$$
$$\mu(1,4) = -[\mu(1,1) + \mu(1,2)] = 0$$
$$\mu(1,6) = -[\mu(1,1) + \mu(1,2) + \mu(1,3)] = +1$$
$$\mu(1,12) = -[\mu(1,1) + \mu(1,2) + \mu(1,3) + \mu(1,4) + \mu(1,6)] = 0$$

Similarly, according to definition (ii), we have

$$\mu(2,1) = \mu(3,1) = \mu(3,2) = \mu(4,1) = \mu(4,2) = \mu(4,3) = \mu(6,1) = \mu(6,2)$$
$$= \mu(6,3) = \mu(6,4) = \mu(12,1) = \mu(12,2) = \mu(12,3) = \mu(12,4)$$
$$= \mu(12,6) = \mu(2,3) = \mu(3,4) = \mu(4,6) = 0$$

Also, based on the recursion it is given that

$$\mu(x,y) \equiv \begin{pmatrix} 1 & -1 & -1 & 0 & 1 & 0 \\ 0 & 1 & 0 & -1 & -1 & 1 \\ 0 & 0 & 1 & 0 & -1 & 0 \\ 0 & 0 & 0 & 1 & 0 & -1 \\ 0 & 0 & 0 & 0 & 1 & -1 \\ 0 & 0 & 0 & 0 & 0 & 1 \end{pmatrix} \quad (A.6)$$

---

[4]Similarly, we have
$$\mu(x,y) = -\sum_{x<z\leq y} \mu(z,y) \quad (A.4)$$

A.5.2 Example

Let $x, y \in \mathbb{P}_f(S)$, that is, both x and y are finite subset of S. Now we define a function

$$\mu(x,y) = \begin{cases} 0 & \text{if } x \not\leq y \\ (-1)^{|y|-|x|} & \text{if } x \leq y \end{cases} \quad (A.7)$$

Obviously, the function $\mu(x,y)$ satisfies conditions (i) and (ii). In fact, $\mu(x,y)$ satisfies condition (iii) too[5].

Let $x \leq y$ and denote $n = y/x$ when $x|y$. Now we can define the binary function $\mu(x,y)$ on P as

$$\mu(x,y) = \begin{cases} \mu(\frac{y}{x}) = \mu(n) & \text{if } x|y \\ 0 & \text{if } x \nmid y \end{cases} \quad (A.8)$$

$\mu(n)$ is just the Möbius function in classical number theory.

A.6 Möbius Inverse Formula on Locally Finite POSET

A.6.1 *Möbius inverse formula A*

Let P be a locally finite POSET with minimum element 0, R represents the real domain, which is a commutative ring with unit element, and $\mu(x,y)$ is Möbius function, which is defined on P and taking value from R. Assuming that both $f(x)$ and $g(x)$ are functions defined on P and taking value from R, then

$$g(x) = \sum_{y \leq x} f(y) \Leftrightarrow f(x) = \sum_{y \leq x} g(y)\mu(y,x) \quad (A.9)$$

Proof. Since P is a locally finite POSET with minimum element 0, interval $[0, x]$ must be finite, and all the summations involved in above formulas

---

[5]Assuming that $x < y$ or $x \subset y$. Also, assuming that $|y| - |x| = r$ and $r > 0$, then for arbitrary $j (0 \leq j \leq r)$ there are totally $\binom{r}{j}$ finite subsets z satisfying the condition $x \leq z \leq y$, and $|z| = |x| + j$. Based on binomial theorem, we have

$$\sum_{x \leq z \leq y} \mu(x,z) = 1 + (-1)\binom{r}{1} + (-1)^2 \binom{r}{2} + \ldots$$

$$+ (-1)^j \binom{r}{j} + (-1)^r \binom{r}{r} = 0$$

Since the conditions (i),(ii) and (iii) are satisfied, the above $\mu(x,y)$ is just the Möbius function on $\mathbb{P}_f(S)$.

must be convergent. Assuming the left part is hold for any $x \in P$, then
$$\sum_{y \leq x} g(y)\mu(y,x) = \sum_{y \leq x}[\sum_{z \leq y} f(z)]\mu(y,x)$$
$$= \sum_{z \leq x} f(z) \sum_{z \leq y \leq x} \mu(y,x)$$
$$= \sum_{z \leq x} f(z)\delta_{z,x} = f(x)$$

Conversely, we have
$$\sum_{y \leq x} f(y) = \sum_{y \leq x}[\sum_{z \leq y} g(z)\mu(z,y)]$$
$$= \sum_{z \leq x} g(z) \sum_{z \leq y \leq x} \mu(z,y)$$
$$= \sum_{z \leq x} g(z)\delta_{z,x} = g(x)$$
$\square$

Assuming that both $f(n)$ and $g(n)$ are defined on $\mathbb{N}$ and taking value from R, and define $d \leq n$ as $d|n$, then
$$g(x) = \sum_{y \leq x} f(y) \Rightarrow g(n) = \sum_{d|n} f(d) \text{ and}$$
$$f(x) = \sum_{y \leq x} g(y)\mu(y,x) \Rightarrow f(n) = \sum_{d|n} \mu(\frac{n}{d})g(d)$$

Therefore, it is given that
$$g(n) = \sum_{d|n} f(d) \Leftrightarrow f(n) = \sum_{d|n} \mu(d)g(\frac{n}{d})$$

vice versa. This is the right Möbius inverse theorem on $\mathbb{N}$. Note that we have gained an insight into this formula by a concept of POSET. Note that inside the set $\mathbb{N}$ there so many subsets being POSET, also, there are so many subsets not being POSET.

A.6.2 Möbius inverse formula B

Let P be a locally finite POSET with maximum element $\mathbf{1}$, R is a commutative ring with unit element, and $\mu(x,y)$ is Möbius function, which is defined on P and taking value from R. Assuming that both $f(x)$ and $g(x)$ are functions defined on P and taking value from R, then
$$g(x) = \sum_{x \leq y} f(y) \Leftrightarrow f(x) = \sum_{x \leq y} \mu(x,y)g(y) \tag{A.10}$$

Proof.
$$\sum_{x\leq y}\mu(x,y)g(y) = \sum_{x\leq y}[\sum_{x\leq y}\mu(x,z)\sum_{y\leq z}f(z)]$$
$$= \sum_{x\leq z}f(z)\sum_{x\leq y\leq z}\mu(x,z)$$
$$= \sum_{x\leq z}f(z)\delta_{z,x} = f(x)$$
□

For example, let $y = nx$ represent $x \leq y$, and
$$\sum_{x\leq y}f(y) \Rightarrow \sum_{n=1}^{\infty}f(nx) \quad \text{with} \quad \sum_{x\leq y}\mu(x,nx)g(y) \Rightarrow \sum_{n=1}^{\infty}\mu(x,nx)g(nx)$$

Set $\mu(x,nx) = \mu(n)$, then the common Möbius series inversion formula is presented.

A.7 Principle of Inclusion and Exclusion

The principle of inclusion and exclusion is one of the most useful methods for counting, and counting is important in physics.

Common treatment in set theory

The elementary idea of the principle can be described as follows.

Assuming that S is a finite full set, and A is a subset of S. Obviously, the number of elements in A is equal to the number of elements in S subtracting the that in the complementary set of A (denoted as $\overline{A}$), that is,

$$|A| = |S| - |\overline{A}| \tag{A.11}$$

or

$$|\overline{A}| = |S| - |A| \tag{A.12}$$

If the elements in A have a certain property P, and the elements in $\overline{A}$ do not have property P, then Eq. (A.12) indicates that the number of elements in S without property P is equal to the number of elements in S subtracting that with property P.

If A_1 and A_2 are subsets of a finite set S, any element in A_1 owns property P_1 and that in A_2 owns property P_2. Now let us count the elements of S, which have neither property P_1, nor property P_2. Obviously, the answer is not $|S| - |A_1| - |A_2|$, because the elements of $A_1 \cap A_2$ have been subtracted twice. Therefore, the answer is

$$|\overline{A_1 \cup A_2}| = |S| - |A_1 \cup A_2| = |A_1| - |A_2| + |A_1 \cap A_2| \tag{A.13}$$

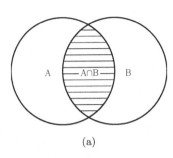
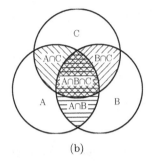

Fig. A.2 (a) A diagram of intersection between A and B. (b) A diagram of intersections among A, B and C.

Again, the number of elements without both properties P_1 and P_2 is equal to the number of elements in S subtracting the number of elements having at least one of the properties or belonging to $A_1 \cup A_2$. Note that if we separate A_1 into two parts: (1) the elements in this part only have property P_1 without property P_2; (2) the elements in this part have not only property P_1 but also property P_2. Obviously, the first part is equal to $A_1 \cup \overline{A_2}$, the second part is $A_1 \cap A_2$. Thus

$$|A_1| = |A_1 \cup \overline{A_2}| + |A_1 \cap A_2| \qquad (A.14)$$

Similarly, we have

$$|A_2| = |A_2 \cup \overline{A_1}| + |A_1 \cap A_2| \qquad (A.15)$$

In the case that A_1, A_2 and A_3 are subsets of S, and we want to count the elements of S, which have neither P_1, nor P_2 or P_3. Then the answer is

$$|\overline{A_1 \cup A_2 \cup A_3}| = |S| - |A_1 \cup A_2 \cup A_3| = |S| - |A_1| - |A_2| - |A_3|$$
$$+ |A_1 \cap A_2| + |A_1 \cap A_3| + |A_2 \cap A_3| - |A_1 \cap A_2 \cap A_3| \qquad (A.16)$$

In this case, A_1 can be separated into three parts: $(1) A_1 \cup \overline{(A_2 \cup A_3)}; (2) A_1 \cap A_2 \cup \overline{A_3}; (3) A_1 \cap A_2 \cap A_3$. In other words, we have

$$|A_1| = |A_1 \cup \overline{(A_2 \cup A_3)}| + |A_1 \cap A_2 \cup \overline{A_3}| + |A_1 \cap A_2 \cap A_3| \qquad (A.17)$$

and so on.

In general, assuming that $\{\mathbb{P}_m\} = P_1, P_2, ..., P_i, ..., P_m$ are some certain kinds of properties of elements of S, and that for any P_i each element of S either owns P_i or does not own it. Now let us count the elements of S, which does not have any of the properties $\{P_i\}$ as follows.

Let $N_i(1 \leq i \leq m)$ represent the number of elements having property P_i, let $N_{ij}(1 \leq i, j \leq m, i \neq j)$ represent the number of elements having both properties P_i and P_j, let $N_{ijk}(1 \leq i, j, k \leq m, and\ i \neq j \neq k,\ with\ i \neq k)$ represent the number of elements having properties P_i, P_j and P_k, and so on. Then the number of elements of S not having any of subsets $P_i(1 \leq i \leq m)$, is equal to[Lin92]

$$\begin{aligned}
|\overline{A_1} \cup \overline{A_2} \cup \cdots \cup \overline{A_m}| &= |\overline{A_1} \cap \overline{A_2} \cap \cdots \cap \overline{A_m}| \\
&= |S| - (|A_1| + |A_2| + \cdots + |A_m|) \\
&\quad + (|A_1 \cup A_2| + |A_1 \cup A_3| + \cdots + |A_{m-1} \cup A_m|) \\
&\quad - (|A_1 \cup A_2 \cup A_3| + \cdots + |A_{m-2} \cup A_{m-1} \cup A_m|) + \cdots \\
&\quad + (-1)^m |A_1 \cup A_2 \cup \cdots \cup A_m|
\end{aligned} \quad (A.18)$$

Sometimes it is written as

$$\begin{aligned}
|\overline{A_1} \cup \overline{A_2} \cup \cdots \cup \overline{A_m}| =& |S| - \{N_1 + N_2 + \cdots + N_m\} \\
& + \{N_{12} + N_{13} + \cdots + N_{m-1,m}\} \\
& + \{N_{123} + \cdots + N_{m-2,m-1,m}\} \\
& + \cdots + (-1)^m N_{1,2,\ldots,m}
\end{aligned}$$

Note that

$$\alpha(1) = \sum_{i=1}^{m} |A_i|$$

$$\alpha(2) = \sum_{i=1}^{m} \sum_{j>i} |A_i \cap A_j|$$

$$\alpha(3) = \sum_{i=1}^{m} \sum_{j>i} \sum_{k>j} |A_i \cap A_j \cap A_k|$$

......

$$\alpha(m) = |A_1 \cap A_2 \cap \ldots \cap A_m| \quad (A.19)$$

Obviously, any element in S having at least one of properties must be counted in α_1, but some elements may be counted repeatedly; any element in S having at least two properties must be counted in α_2, but some elements may be counted repeatedly. Similarly, any element in S having at least j properties must be counted in α_j, with some elements counted repeatedly.

Now let us introduce β_i to represent the number of elements just having

i (neither more nor less than i) from m properties. This can be given as follows.

$$\beta(0) = |\overline{A_1} \cap \overline{A_2} \cap \cdots \cap \overline{A_m}|$$

$$\beta(1) = \sum_{i=1}^{m} |\overline{A_1} \cap \overline{A_2} \cap \cdots \cap \overline{A_{i-1}} \cap A_i \cap \overline{A_{i+1}} \cap \cdots \cap \overline{A_m}|$$

$$\beta(2) = \sum_{i=1}^{m} \sum_{j>i}^{m} |\overline{A_1} \cap \overline{A_2} \cap \cdots \cap \overline{A_{i-1}} \cap A_i \cap \overline{A_{i+1}} \cap \cdots$$
$$\cap \overline{A_{j-1}} \cap A_j \cap \overline{A_{j+1}} \cap \cdots \cap \overline{A_m}|$$

$$\beta(m) = |\overline{A_1} \cap \overline{A_2} \cap \ldots \cap \overline{A_m}| \qquad (A.20)$$

According to the definition of α_i and β_i, there is a relation between them as follows

$$\beta(k) = \alpha(k) - \binom{k+1}{1}\alpha(k+1) + \binom{k+2}{2}\alpha(k+2) - \cdots + (-1)^{m-k}\binom{m}{m-k}\alpha(m)$$

Proof. For any element $x \in S$, there are three cases.

Case 1. The number of properties element x has is less than k, then x does not contribute anything to β_k and $\alpha(k), \alpha(k+1), ...$, both sides of the above equation are zero.

Case 2. The number of properties element x has is just equal to k, then x is counted once in $\beta(k)$, once in $\alpha(k)$, and x has not been counted in $\alpha(k+i)$ when $i > 0$. Thus both sides of the above equation are equal to 1.

Case 3. The number of properties element x has is more than k. Let the number of properties x has is equal to $k+i$ with $1 \leq i \leq m-k$. Thus, x is not counted in β_k, and x is not counted in $\alpha(k+j)$ too when $j > i$.

Since x has $k+i$ properties, x must be counted $\binom{k+i}{k}$ times in $\alpha(k)$, $\binom{k+i}{k+1}$ times in $\alpha(k+1), \cdot, \binom{k+i}{k+i} = 1$ times in $\alpha(k+i)$. Denoting the right side of above equation as R, then

$$R = \alpha(k) - \binom{k+1}{1}\alpha(k+1) + \binom{k+2}{2}\alpha(k+2) - \cdots$$
$$+ (-1)^{m-k}\binom{m}{m-k}\alpha(m)$$
$$= \binom{k+0}{0}\binom{k+i}{k} - \binom{k+1}{1}\binom{k+i}{k+1} + \binom{k+2}{2}\binom{k+i}{k+2}$$
$$- \cdots + (-1)^i \binom{k+i}{i}\binom{k+i}{k+i}$$

Note that

$$\binom{k+t}{k}\binom{k+i}{k+t} = \frac{(k+t)!}{t!k!} \cdot \frac{(k+i)!}{(k+t)!(i-t)!}$$
$$= \frac{(k+i)!}{k!i!} \cdot \frac{i!}{t!}(i-t)! = \binom{k+i}{k}\binom{i}{t}$$

it is given that

$$R = \binom{k+i}{k}\binom{i}{0} - \binom{k+i}{k}\binom{i}{1} + \ldots + (-1)^i \binom{k+i}{k}\binom{i}{i}$$
$$= \binom{k+i}{k}\left\{\binom{i}{0} - \binom{i}{1} + \ldots + (-1)^i \binom{i}{i}\right\} = 0$$

From the above, it is proven. □

Note that by using these notations, Eq. (A.18) can be written as

$$\beta_m = |\overline{A_1} \cap \overline{A_2} \cap \ldots \cap \overline{A_m}| = |S| - \sum_{i=1}^{m}(-1)^{i+1}\alpha_i$$

and considering that

$$|\overline{A_1} \cap \overline{A_2} \cap \ldots \cap \overline{A_m}| = |S| - |A_1 \cup A_2 \cup \ldots \cup A_m|$$

it is given that

$$|A_1 \cup A_2 \cup \ldots \cup A_m| = \sum_{i=1}^{m}(-1)^{i+1}\alpha_i$$

A.8 Cluster Expansion Method

First of all, let us consider a system with five atoms. In principle, there are 2-body, 3-body, 4-body and 5-body interactions between these atoms[Dra2004; Sun2008]. Assuming that only 2-body, 3-body and 4-body interactions are involved, then there are $\binom{5}{4} = 5$ four-particle clusters, $\binom{5}{3} = 10$ three-particle clusters and $\binom{5}{2} = 10$ two-particle pairs. Now let us consider a 4-particle cluster with configuration (X_1, X_2, X_3, X_4). Within this cluster, there are only one four-particle clusters, $\binom{4}{3} = 4$ three-particle clusters and $\binom{4}{2} = 6$ two-particle pairs. Define the interaction among atoms within a cluster as the total energy of the cluster subtracting the sum of energies of all isolated atoms, and assuming that the interacting potentials are only dependent on the positions and species of all atoms within the cluster and independent of environment, then we have

$$\begin{cases} V(1,2) = E(1,2) - E(1) - E(2) = \mathcal{E}(1,2) \\ V(1,3) = E(1,3) - E(1) - E(3) = \mathcal{E}(1,3) \\ V(1,4) = E(1,4) - E(1) - E(4) = \mathcal{E}(1,4) \\ V(2,3) = E(2,3) - E(2) - E(3) = \mathcal{E}(2,3) \\ V(2,4) = E(2,4) - E(2) - E(4) = \mathcal{E}(2,4) \\ V(3,4) = E(3,4) - E(3) - E(4) = \mathcal{E}(3,4) \\ V(1,2,3) + V(1,2) + V(1,3) + V(2,3) = \\ \quad = E(1,2,3) - E(1) - E(2) - E(3) = \mathcal{E}(1,2,3) \\ V(1,2,4) + V(1,2) + V(1,4) + V(2,4) = \mathcal{E}(1,2,4) \\ V(1,3,4) + V(1,3) + V(1,4) + V(3,4) = \mathcal{E}(1,3,4) \\ V(2,3,4) + V(2,3) + V(2,4) + V(3,4) = \mathcal{E}(2,3,4) \\ V(1,2,3,4) + V(1,2,3) + V(1,2,4) + V(1,3,4) + V(2,3,4) \\ \quad + V(1,2) + V(1,3) + V(1,4) + V(2,3) + V(2,4) + V(3,4) \\ \quad = E(1,2,3,4) - E(1) - E(2) - E(3) - E(4) \\ \quad = \mathcal{E}(1,2,3,4) \end{cases} \quad \text{(A.21)}$$

where the variable $(1,2,3)$ represents (X_1, X_2, X_3), and so on. Note that

$\mathcal{E}(1,2,3,4), \mathcal{E}(1,2,3), \mathcal{E}(1,2,4), \mathcal{E}(1,3,4), \mathcal{E}(2,3,4), \mathcal{E}(1,2), \mathcal{E}(1,3), \mathcal{E}(1,4),$
$\mathcal{E}(2,3), \mathcal{E}(2,4), \mathcal{E}(3,4)$ can be calculated by using *ab initio* calculation

method. Assuming the interaction potentials are only dependent on atomic species and positions, and independent to environment of the atomic clusters, the above set of equations has unique solution as $V(1,2) = \mathcal{E}(1,2), V(1,3) = \mathcal{E}(1,3), V(1,4) = \mathcal{E}(1,4), V(2,3) = \mathcal{E}(2,3), V(2,4) = \mathcal{E}(2,4)$ and $V(3,4) = \mathcal{E}(3,4)$, and

$$\begin{cases} V(1,2,3) = & \mathcal{E}(1,2,3) - \{\mathcal{E}(1,2) + \mathcal{E}(1,3) + \mathcal{E}(2,3)\} \\ V(1,2,4) = & \mathcal{E}(1,2,4) - \{\mathcal{E}(1,2) + \mathcal{E}(1,4) + \mathcal{E}(2,4)\} \\ V(1,3,4) = & \mathcal{E}(1,3,4) - \{\mathcal{E}(1,3) + \mathcal{E}(1,4) + \mathcal{E}(3,4)\} \\ V(2,3,4) = & \mathcal{E}(2,3,4) - \{\mathcal{E}(2,3) + \mathcal{E}(2,4) + \mathcal{E}(3,4)\} \\ V(1,2,3,4) = & \mathcal{E}(1,2,3,4) - \{\mathcal{E}(1,2,3) + \mathcal{E}(1,2,4) \\ & + \mathcal{E}(1,3,4) + \mathcal{E}(2,3,4)\} + \{\mathcal{E}(1,2) + \mathcal{E}(1,3) \\ & + \mathcal{E}(1,4) + \mathcal{E}(2,3) + \mathcal{E}(2,4) + \mathcal{E}(3,4)\} \end{cases} \quad \text{(A.22)}$$

Similarly, it is given that
$$\begin{cases} V(1,2) = \mathcal{E}(1,2) \\ V(1,3) = \mathcal{E}(1,3) \\ V(1,5) = \mathcal{E}(1,5) \\ V(2,3) = \mathcal{E}(2,3) \\ V(2,5) = \mathcal{E}(2,5) \\ V(3,5) = \mathcal{E}(3,5) \\ V(1,2,3) = \mathcal{E}(1,2,3) - \{\mathcal{E}(1,2) + \mathcal{E}(1,3) + \mathcal{E}(2,3)\} \\ V(1,2,5) = \mathcal{E}(1,2,5) - \{\mathcal{E}(1,2) + \mathcal{E}(1,5) + \mathcal{E}(2,5)\} \\ V(1,3,5) = \mathcal{E}(1,3,5) - \{\mathcal{E}(1,3) + \mathcal{E}(1,5) + \mathcal{E}(3,5)\} \\ V(2,3,5) = \mathcal{E}(2,3,5) - \{\mathcal{E}(2,3) + \mathcal{E}(2,5) + \mathcal{E}(3,5)\} \\ V(1,2,3,5) = \mathcal{E}(1,2,3,5) - \{\mathcal{E}(1,2,3) + \mathcal{E}(1,2,5) \\ \qquad\qquad\quad + \mathcal{E}(1,3,5) + \mathcal{E}(2,3,5)\}\{\mathcal{E}(1,2) + \mathcal{E}(1,3) \\ \qquad\qquad\quad + \mathcal{E}(1,5) + \mathcal{E}(2,3) + \mathcal{E}(2,5) + \mathcal{E}(3,5)\} \end{cases} \quad (A.23)$$

By using the language in POSET, we are facing a finite POSET $P_f(S)$ with all the clusters as its elements:

$$(1,2), (1,3), ..., (4,5), \text{ and } (1,2,3), (1,2,4), ..., (3,4,5),$$
$$(1,2,3,4), (1,2,3,5), ..., (2,3,4,5) \text{ and } (1,2,3,4,5)$$

Then we can define a series of partially ordered binary relations between them, such as

$$(1,2) \subseteq (1,2,3) \subseteq (1,2,3,4) \subseteq (1,2,3,4,5)$$
$$\Rightarrow (1,2) \leq (1,2,3) \leq (1,2,3,4) \leq (1,2,3,4,5)$$

and

$$(1,2) \subseteq (1,2,4) \subseteq (1,2,4,5) \subseteq (1,2,3,4,5)$$
$$\Rightarrow (1,2) \leq (1,2,4) \leq (1,2,4,5) \leq (1,2,3,4,5)$$

Based on these notations, the relations between cohesive energy of any cluster and interatomic potentials within the cluster[6] can be given that

$$\mathcal{E}(x) = \sum_{y \leq x} V(y) \qquad (A.24)$$

for any $x, y \in P_f(S)$, the corresponding Möbius inverse formula can be expressed as

$$V(x) = \sum_{y \leq x} \mathcal{E}(y) \mu(y, x) \qquad (A.25)$$

---

[6] Consider the difference with that in crystal.

where the Möbius function is defined as

$$\mu(x, y) = \begin{cases} 0 & \text{if } y \not\leq x \\ (-1)^{|x|-|y|}, & \text{if } y \leq x \end{cases} \quad (A.26)$$

In general, we denote the total energy of a M particle system as

$$E_P = E_P(X_1, X_2, ..., X_M) \quad (A.27)$$

and the corresponding cohesive energy is denoted as

$$E_P = E_P(X_1, X_2, ..., X_M) - \sum_{n=1}^{M} E(X_n) \quad (A.28)$$

Note that we set P as the order of the expansion, in other words, $N = 2, ..., P$ denotes N-atom cluster, all the $V^{(P+s)}(s \geq 1)$ will be ignored. Now let us represent the cohesive energy $\mathcal{E}_P$ as a sum of cohesive energies of all possible clusters within the M-atom system as

$$\mathcal{E}_P = \mathcal{E}_P(X_1, X_2, ..., X_M) = \sum_{N=2}^{P} \mathcal{E}^{(N)}(X_1, X_2, ..., X_M) \quad (A.29)$$

and the energy of each cluster can be represented as a summation of N-body interaction potentials $V^{(N)}$ via

$$\mathcal{E}^{(N)}(X_1, X_2, ..., X_M) = \sum_{m_1=1}^{M} \sum_{m_2=m_1+1}^{M} ... \sum_{m_N=m_{N-1}+1}^{M}$$
$$V^{(N)}(X_{m_1}, X_{m_2}, ..., X_{m_M}) \quad (A.30)$$

where the summation runs over all the pairwise different indices. It is generally assumed that the potentials $V^{(N)}$ can be chosen as being independent of the environment in which the atoms $\mathbf{R}_1, \mathbf{R}_2, ..., \mathbf{R}_N$ are embedded, so that they are structure independent and therefore transferable to any atomic configuration, including all solid and liquid configurations formed by the atoms. The inversion of Eq. (A.30) is straightforward for small values of M[2]. By Möbius inverse formula in POSET, it is given that

$$V^{(N)}(X_1, X_2, ..., X_N) = \sum_{L=2}^{N} (-1)^{N-L} \sum_{m_1=1}^{M} \sum_{m_2=m_1+1}^{M} ...$$
$$\sum_{m_L=m_{L-1}+1}^{M} \mathcal{E}_L(X_{m_1}, X_{m_2}, ..., X_{m_L}) \quad (A.31)$$

Note that the above deduction constitutes a unique definition of N-body potentials $V^{(N)}$ which are structure independent because this equation does not carry any information about the environment of the atom clusters. Once the potentials $V(N)$ have been constructed, they can be used to calculate the energy $\mathcal{E}(M)$ and E(M) according to above equations for any gaseous, liquid or solid state of the material, i.e., these potentials $V(N)$ are perfectly transferable by construction. Of course, in the calculation for a realistic system there may be other problems such as the increasing total number of clusters involved in the cluster expansion for large M and P. In practice, it is important to design an approximation to reduce the total number of clusters, then we can use a simple way for inversion. Note that building the N-body potentials from both *ab initio* and atomistic calculations is still an interesting and challenging problem[Sun2008]. Finally, remember, there is a well-known saying from Einstein that as far as the laws of mathematics refer to reality, they are not certain, and as far as they are certain, they do not refer to reality[Pap95].

Epilogue

Derived in 1832 by Möbius, the classical Möbius inverse formula in Number Theory has since become well known in the mathematical community. However, one might observe that there has been a wide gap between the bare statement of principle and the skill for physical application. After more than 150 years in which the Möbius formula has been considered as purely academic, or beyond what was useful, in the physics community, this apparently obscure result in classical mathematics suddenly appears to be connected with a variety of important inverse problems in the physical sciences.

Even though the original work on series inversion was imperfect, the creativity of Möbius' work are unquestionably distinctive. Thus this insufficient work immediately led to the Möbius inverse formula in number theory, and the Möbius inversion had been generally acknowledged as concise, ripe and perfect, so that it has long held an unchallenged subject in mathematical field. In order to apply this pure mathematical result to physics, much imperfectness appears and shows renewed vitality: many new results with a lot of new problems have been emerging in an endless stream.

This book provided a concise application-oriented introduction to Möbius inversion formulas with many applications to wide disciplines in the physical sciences. A few examples by way of review: The application to boson (or fermion) system is set to designing Taylor's expansion of the Bose (or Fermi) distribution, with corresponding semigroup structure. The application to interatomic potentials, whether for bulk materials or interfaces, is based on designing special energy processes or curves for relatively structurally simple models, which can be performed by *ab initio* calculations, from which the required interatomic potentials are then extracted by the

Möbius inversion method. The mechanical and thermodynamic properties of more complex structures can then be evaluated by these potentials. Also, the Möbius inversion formula on partially ordered sets[Rot64] and the corresponding applications such as the cluster method variation (CVM) and cluster expansion method (CEM) have been developing quickly. This brief introductory book is quite far from the end of the story. Broad horizons of emerging applications remain to be explored.

It should be expected that in the quantum age, the number-theoretic technique will become an honorable servant to the science and technology community. If this book has built a bridge between this branch of pure mathematics and concrete physical problems, others, with basic knowledge and practical wisdom in the applied physics, we must design the vehicle to cross it. "Where will all this lead? The idea would be that somebody should put a previously unsolved problem through the new Möbius mill"[Mad90].

Finally, the author should feel happy and shame. Why happy? Because there are so many branches of physics connecting with Möbius inversion. Why shame? Because we still do not know too much knowledge on convergence and stability in Möbius series inversion, and ill-posedness in a general inverse problem. Fortunately, as Hilbert said, "As long as a branch of science offers an abundance of problems, so long is it alive." Now let us quote one sentence further from Feynman as an excuse or fig leaf: "People have a habit in writing articles published in scientific journals to make the work as finished as possible, to cover up all the tracks, to not worry about the blind alleys or describe how you had the wrong idea first, and so on. So there isn't any place to publish, in a dignified manner, what you actually did in order to get to do the work"[Fey66]. To switch, a bit abruptly, to a metaphor, the author's great hope is to offer banal remarks to spark abler talk by others.

David Hilbert(1862–1943)

Richard Feynman (1918–1988)

Bibliography

[An88] G.Z. An, A note on the Cluster Variation Method, J.Stat.Phys., 52 (1988) 727

[Apo76] T.M. Apostol, Introduction to analytic number theory, Springer, Berlin (1976) Ch.12

[Ram2000] J.P.P. Ramalho, G.V. Smirnov, On the structure of a local isotherm and solution to the adsorption integral equation, Langmuir, 16 (2000) 1918

[Bas99] M.I. Baskes, Many-body effects in fcc metals: a Lennard-Jones embedded atom potentials, PRL 83 (1999)2592

[Bas2000] B.J. Lee and M.I. Baskes, Second nearest-neighbor modified embedded atom method potential, Phys.Rev. B62 (2000) 8564

[Bas2007] M.I. Baskes, Multistate modified embedded atom method potential, Phys.Rev. B75 (2007) 094113

[Baz96] M.Z. Bazant and E. Kaxiras, Modeling of covalent bonding in solids by inversion of cohesive energy curves, Phys.Rev.Lett., 77 (1996) 4370

[Baz97] M.Z. Bazant, E. Kaxiras and J.F. Justo, Environment-dependent interatomic potential for bulk sillicon, Phys.Rev. B56 (1997) 8542

[Baz97-1] M.Z. Bazant, Interatomic forces in covalent solids, PhD Thesis, Harvard University, (1997)

[Baz98] M.Z. Bazant, Mobius Series Inversion Rediscoved, homepage, Harvard (1998)

[Baz99] M.Z. Bazant, Historical Notes, MIT IPP Mathematics Lecture, 1999

[Bea75] K.G. Beauchamp, Walsh functions and their applications, Academic Press, New York, 1975

[Bel66] R. Bellman, R.E. Kalaba and J.A. Lockett, Numerical Inversion of the Laplace Transform, Elsevier, Amsterdam, 1966

[Bet2007] T.J. Bethell,A. Chepurnov,A. Lazarian and J. Kim, Polarization of dust emission in clump molecular clouds and cores, Astrophys Journal, 663 (2007) 1055Y1068

[Bev89] R.M. Bevensee, Comments on recent solutions to the inverse black body radiation problem , IEEE Trans. AP-37 (1989) 1635

[Bog59] N.N. Bogoliubov, Introduction to the Theory of Quantized Fields, Interscience Publisher, New York, 1959

[Boj82] N.N. Bojarski, Inverse Black Body Radiation, IEEE Trans.Antennas Propagat., AP-30 (1982) 778

[Boj84] N.N. Bojarski, Closed Form Approximations to the Inverse Black Body Radiation Problem, IEEE Trans.Antennas Propagat., AP-32 (1984) 415

[Bor2001] G.L. Boreman, Modulation transfer function in optical and electro-optical systems, SPIE Press,Washington, 2001

[Bru03] H. Bruns, Grundlinien des Wissenschaftlichichnen Rechnens, B.G. Teubner, Verlag, Leipzig, 1903

[Bur91] S.A. Burr(Ed.), The Unreasonable Effectiveness of Number Theory, Proceedings of Symposia in Applied Mathematics, Volume 46, New York, 1991

[Bus2001] V.A. Bushenkov, J.P.P. Ramalho, G.V. Smirnov, Adsorption Integral Equation via Complex Approximation with Constraints, J.Computational Chemistry, 22 (2001) 1058

[Cai2005] J. Cai, X.Y. Hu and N.X. Chen, Multiple lattice inversion approch to interatomic potentials for compound semiconductors, J.Phys.Chem.Solid., 66 (2005) 1256

[Cai2007a] J. Cai, N.X. Chen and H.Y. Huang, Atomistic study of the pressure-induced phase-transition mechanism in GaAs by Mobius inversion potentials, J.Phys.Chem.Solid. 68 (2007) 445

[Cai2007b] J. Cai and N.X. Chen,Phys.Rev.B75 (2007) 134109

[Cai2007c] J. Cai and N.X. Chen, Theoretical study of pressure-induced phase transition in AlAs: from zinc-blende to NiAs structure, Phys.Rev. B75 (2007) 174116

[Car80] A.E.Carlsson, C.D.Gelatt and H.Ehrenreich, An ab initio pair potential applied to metals, Philos., Mag., A 41 (1980) 241

[Car90] A.E.Carsson, Beyond pair potentials in elemental transition metals and semiconductors, in Solid State Physics: Advances in Research and Applications, edited by H.Ehrenreich and D.Turnbull (Academic, New York, 1990), 43, pp.1-91

[Cer80] G.F.Cerofolini, Questions of method in a large class of improperly-posed problems, 1980

[Ces1885] E.Cesàro, Sur l'inversion de certains series, Annali di Mathematiche Pura ed Applicata (2) 13,339-351(1885)

[Cha61] R.G. Chambers, The inversion of specific heat curves, Proc.Phys.Soc.,78 (1961) 941

[Che96] N.X. Chen, Z.D. Chen, S.J. Liu, Y.N. Shen, X.J. Ge, Algebraic rings of integers and some 2D lattice problems in physics, J.Phys. A29 (1996) 5591

[Che2008] Y. Chen, Interatomic potential: theory, application, and challenge, Session proposal for Codata 2008.

[Cha73] D.C. Champeney, Fourier transforms and their physical applications, Academic Press, London, 1973.

[Che90] N.X. Chen, Modified Mobius inverse formula and its applications in physics, Phys.Rev.Lett. 64(1990)1193

[Che90a] N.X. Chen and G.Y. Li, Theoretical investigation on the inverse black-body radiation problem, IEEE Trans. AP-38 (1990) 1287

[Che91] N.X. Chen and G.B. Ren, Inverse problems on Fermi systems and ionic crystals, Phys.Lett. A160 (1991) 319

[Che93] N.X. Chen, C.F. Zhang, M. Zhou, G.B. Ren and W.B. Zhao, Closed-form solution for inverse problems of Fermi systems, Phys.Rev. E48 (1993) 1558

[Che94] N.X. Chen, Z.D. Chen, Y.N. Shen, 3D Inverse lattice problems and Möbius inversion, 184, (1994) 347

[Che96] N.X. Chen, Z.D. Chen, S.J. Liu, Y.N. Shen, X.J. Ge, Algebraic rings of integers and some 2D lattice problems in physics,J.Phys. A29 (1996) 5591

[Che97] N.X. Chen, Z.D. Chen, Y.C. Wei, Multi-dimensional lattice inversion problem and a uniformly sampled arithmetic Fourier transform, Phys.Rev. E55 (1997) R5

[Che98] N.X. Chen and E.Q. Rong, Unified solution of the inverse capacity problem, Phys.Rev. E57 (1998) 1302; 6216

[Che98] N.X. Chen, X.J. Ge, W.Q. Zhang and F.W. Zhu, Atomistic analysis of the field-ion microscopy image of Fe_3Al, Phys.Rev. B57 (1998) 14203

[Che98z] Z.D. Chen, Y.N. Shen, S.J. Liu, Mobius inversion on a multiplicative semigroup and inverse lattice problems in physics, Math. Methods in Appl. Sci., 21 (1998)269

[Che1851] P.L. Chebyshev, Note Sur differentes *séries*, Journale de *Mathématiques* Pures et *Appliquées*, ed. by Liouville, (1) 16(1851)

[Che2001] N.X. Chen, J. Shen and X.P. Su, Theoretical study on the phase stability, site preference, and lattice parameters for Gd(Fe,T)12, J.Phys.-cond.matt. 13 (2001) 2727

[Che2001z] Z.D. Chen, Y.N. Shen, J.Ding, The Mobius inversion and Fourier coefficients, Appl.Math. and Comput., 17 (2001) 161

[Che2008] Y. Chen, Interatomic potential: theory, application, and challenge, Private communication, Codata, 2008.

[Col54] J. Coltman, The specification of imaging properties by response to a sine wave input, JOSA 44(1954)468

[Coo50] R.G. Cooke, Infinite matrices and sequence spaces, MacMillan and Co., Limited, London, 1950

[Coo65] J.W. Cooley and J.W.Tukey, An algorithm for the machine calculation of complex Fourier Series, Math.Comput. 19 (1965) 297

[Dai87] D.S. Dai and C.M. Qian, Ferromagnetism, Vol.1, Ch.4, Science Publishing House, Beijing, 1987

[Dai90] X.X. Dai, X.W. Xu and J.Q. Dai, Phys.Lett. A 147 (1990) 445

[Dav81] P.J. Davis and R. Hersh, The mathematical experience, Houghton Mifflin Com. (1981) Boston

[Daw83] M.S. Daw and M.I. BaskesSemiempirical Quantum Mechanical Calculation of Hydrogen Embrittlement in Metal, PRL50(1983)1285

[Daw84] M.S. Daw and M.I. BaskesEmbedded Atom Method: Drivation and application to impurity, surface, and other defect s in metals,PRB29(1984)6443

[Deb12] P. Debye, Zur Theorie der spezifischen Wärmen, Ann. d. Phys., 39 (1912) 789

[Din2000] X.Q. Ding and P.Z. Luo, Weak Convergence of Some series, Acta Math. Science, 20B (2000) 433
[Dou92a] L. Dou and R.J. Hodgson, Maximum entropy method in inverse blackbody radiation problem, J.Appl.Phys., 71(1992) 3159
[Dou92] L. Dou and R.J. Hodgson, Application of the regularization method to the inverse blackbody radiation problem, IEEE Trans. AP-40 (1992) 1249
[Dra2004] R. Drautz, M. Fahnle and J.M. Sanchez, General relations between many-body potentials and cluster expansions in multicomponent systems, J.Phys:Cond.Matt. 16 (2004)3843
[Ein07] A. Einstein, Die Plancksche Theorie der Strahlung und die Theorie der spezifischen Wärme, Ann. d. Phys., **22** (1907) 180
[Esp80] E. Esposito, A.E. Carlsson, D.D. Ling, H. Ehrenreich and C.D. Gelatt, First-prinnciples calculations of the theoretical tensile strength of copper, Philos., Mag., A 41 (1980) 251
[Ewa21] P.P. Ewald, Die Berchnung optischer und elektrostatischer Gitterpotentiale, Ann.Phys. 64 (1921) 253
[Fey66] R. Feynman, 1966 Nobel Lecture
[Fey67] R. Feynman, The Character of physical law, MIT Press, 1967
[Fin69] L. Finegold and N.Phillips, Low Temperature Heat Capacities of Solid Argon and Krypton, Phys.Rev. 177 (1969) 1383
[Fro58] H. Fröhlich, Theory of Dielectrics, 2nd. Clarendon (1958) Oxford
[Ge97] X.J. Ge, N.X. Chen, Z.D. Chen, Efficient Algorithm for 2-D Arithmetic Fourier Transform, IEEE TRANS.SIGNAL PROCESSING, 45 (1997) 2136
[Ge99] X.J. Ge, N.X. Chen, W.Q. Zhang and F.W.Zhu, Selective field evaporation in field-ion microscopy for ordered alloys, J.Appl.Phys. 85 (1999) 3488
[Gio70] A.A. Gioia, The theory of numbers, Dover, New York, 1970
[Gos61] N. Gossard, Phys.Rev.Lett. 7 (1961) 122
[Gra80] I.S. Gradshteyn and I. M. Ryzhik, Table of Integrals, Series, and Products Academic, New York, 1980
[Gro99] C.W. Groetsh, Inverse Problems– Activities for Undergraduates, The Mathematical Association, 1999
[Had23] J. Hadamard, Lecture on the Cauchy Problem in Linear Partial Differential Equations. Yale Univ. Press, New Haven, 1923
[Ham83] M. Hamid and H.A. Ragheb, Inverse blackbody radiation at microwave frequencies, IEEE Trans.AP-31 (1983) 810
[Hao93] B.L. Hao, Introduction to chaos dynamics (from parabolic), Shanghai Press of Science and Technology, Shanghai, 1993
[Hao2002] S.J. Hao, N.X. Chen and J. Shen, The phase group of $Nd3Fe29 - xTix$(A2/m or P2-1/C), Phys.stat.sol. B 234 (2002) 487
[Hou78] W.A. House, Adsorption on a random configuration of adsorptive hetrogeneities, J. Colloid Interface Sci., 67 (1978) 166
[Hsu94] C.C. Hsu, I.S. Reed, and T.K. Truong, Inverse Z-Transform by Möbius Inversion and Error Bounds of Aliasing in Sampling, IEEE Trans. Signal Processing, 42 (1994) 2823
[Hug90] B.D. Hughes, N.E. Frankel and B.W. Ninham, Chen's inversion formula, Phys.Rev. A42 (1990) 643

[Hug97] B.D. Hughes, Some applications of classical analysis in physics and physical chemistry, Colloids and Surfaces, A129 (1997) 185

[Hun86] J.D. Hunter, An Improved Closed-Form Approximation to the Inverse Black Body Radiation Problem at Microwave Frequencies, IEEE TRANS. AP 34 (1986) 261

[Ji2006] F.M. Ji, J.P. Ye, L. Sun, X.X. Dai et al., An inverse transmissivity problem, its Mobius inversion solusion and new practical solution method, Phys.Lett. A 352 (2006)467

[Kan2002] Y.M. Kang, N.X.C hen and J. Shen, Site preference and vibrational properties of Ni_3Al with ternary additions Pd and Ag, Modern Phys. Lett. B16 (2002)727

[Kel93] B.T. Kelley and V.K. Madisetti, Efficient VLSI architecture for the arithmetic Fourier transform(AFT), IEEE Trans. Signal Processing, 41 (1993) 365

[Kik51] R. Kikuchi, A theory of cooperative phenomena, Phys.Rev.81 (1951)988

[Kim85] Y. Kim and D.L. Jaggard, Inverse Black Body Radiation: An Exact Closed-Form Solution, IEEE TRANS. AP-33, (1985) 797

[Kim2006] Y.M. Kim, and B-J.Lee, Modified embedded-atom method potential for Ti and Zr, Phys.Rev. B74 (2006) 014101

[Kir96] A. Kirsch, An introduction to the mathematical theorem of inverse problems, Springer-Verlag, 1996, New York

[Kit76] C. Kittel, Introduction to Solid State Physics, Wiley-Sons, New York, 1976

[Kno28] K. Knopp, Theory and application of infinite series, English edition, 928

[Kra87] H.J. Krause, J.E.Wittig and G.Frommeyer, A comparative study of B2-DO3 ordered iron-aluminium alloys using atom probe-field ion microscopy and transmission electron microscopy, Z.Metalkd., 78(1987) 576

[Lak84] N. Lakhtakia and A. Lakhtakia, Inverse Black Body Radiation at Submillimeter Wavelengths, IEEE TRANS, AP-32 (1984) 872

[Lan76] U. Landman and E.W. Montroll, Adsorption on heterogeneous surfaces. I. Evaluation of the energy distribution function via the Wiener and Hopf method, J.Chem.Phys. 64 (1976) 1762

[Las2000] R. Laskowski, Double lattice inversion technique - application to the EAM potential construction, Phys.Stat.Sol. (b)222 (2000) 457

[Leb2004] P. Leboeuf, Periodic orbit spectrum in terms of Ruelle-Pollicott resonances, Phys.Rev. E 69(2004) 026204

[Lee2000] B-J. Lee and M.I. Baskes, Second nearest-neighbor modified embedded-atom method potential , Phys.Rev.B62 (2000)8564

[Lee2001] B-J. Lee, M.I. Baskes, H.C. Kim, and Y.K. Cho, Second nearest-neighbor modified embedded-atom method potential for bcc transition metals, Phys.Rev. B64 (2001)184102

[Lee2003] B-J. Lee, J-H. Shim and M.I. Baskes, Semiempirical atomic potentials for the fcc metals Cu, Ag, Au, Ni, Pd, Pt, Al, and Pb based on first and second nearest-neighbor modified embedded-atom method , Phys.Rev. B68 (2003)144112

[Li99] D. Li, P.F. Goldsmith and T.L. Xie, A new method for determining the dust temperature distribution in star-forming regions,AstrophysJ.522(1999) 897

[Li2005] H.Y. Li, Solution of Inverse Blackbody Radiation Problem with Conjugation Gradient Method, IEEE Trans. AP-53 (2005) 1840

[Lif54] I.M. Lifshitz, Determination of energy spectrum of a boson system by its heat capacity, Zh.Eksp.Theor.Fiz. 26 (1954) 551

[Lig91] V.A. Ligachev and V.A. Filikov, A new method for calculating relaxation time spectra, and its application to the study of α- Si:H, Fiz.Tverd.Tela. 33 (1991) 3292 [Sov.Phys.Solid State **33**(1991) 1857]

[Lin92] J.H. van Lint and R.M. Wilson, A course in combinatorics, 2nd ed, 1992, Cambridge University Press

[Lon2005a] Y. Long, N.X. Chen and W.Q. Zhang, Pair potentials for a metalCceramic interface by inversion of adhesive energy, J.Phys: Cond. Matt. 17 (2005) 2045

[Lon2005b] Y. Long, N.X. Chen and H.Y. Wang, Theoretical investigations of misfit dislocations in Pd/MgO(001) interfaces,J. Phys.: Cond. Matt 17 (2005) 6149

[Lon2007] Y. Long and N.X. Chen, Pair potential approach for metal/Al2O3 interface, J.Phys.: Cond.Matt. 19 (2007) 196216

[Lon2008a] Y. Long and N.X. Chen, Atomistic study of metal clusters supported on oxide surface, Surface Science 602 (2008) 46

[Lon2008b] Y. Long and N.X. Chen, Interface reconstruction and dislocation networks for a metal/alumina interface: an atomistic approach, J. Phys.:Condens. Matter 20 (2008) 135005

[Lon2008c] Y. Long and N.X. Chen, An atomistic simulation and phenomenological approach of misfit dislocation in metal/oxide interfaces, Surface Science 602 (2008) 1122

[Lon2008d] Y. Long and N.X. Chen, Atomistic simulation of misfit dislocation in metal/oxide interface, Computational Materials Science, 42 (2008) 426

[Lon2008e] Y. Long and N.X. Chen, Atomistic study of metal clusters supported on oxide surface, Surface Science 602 (2008) 46

[Lon2008f] Y. Long and N.X. Chen, Molecular dynamics simulation of Pd clusters colliding onto Mg(001) surface, Physica B403 (2008) 4006

[Lon2008g] Y. Long and N.X. Chen, A theoretical exploration of strain-mediated interaction for supported nanoclusters, Surface Science 602 (2008) 3408

[Lon2009] Y. Long and N.X. Chen, Theoretical study of (Ag, Au, Cu)/Al$_2$O$_3$ interfaces, J. Phys.:Condens. Matter 21 (2009) 315003

[Lor86] J.W. Loram, On the determination of the phonon density of states from the specific heat, J.Phys.C: Solid State Phys., 19 (1986) 6113

[Luo99] X. Luo, G.F. Qian, E.G. Wang and C.F. Chen, Molecular-dynamics simulation of Al/SiC interface structure, Phys. Rev. B 59 (1999) 10125

[Mad90] J. Maddox, Mobius and problems of inversion, Nature 344(1990) 377

[Mai79] W.L. Mei, Optical transfer function and its foundation in mathematics and physics, Defence Press, Beijing, 1979

[Mer68] N.D. Mermin, Crystalline order in two dimensions, Phys.Rev., 176 (1968) 250

[Mil89] M.K. Miller and C.D.W. Smith, Atom Probe Microanalysis, Materials Research Society, Pittsburgh, 1989
[Mis70] D.N. Misra, New Adsorption Isotherm for hetrogeneous Surface, J.Chen.Phys., 52(1970) 5499
[Mon42] E.W. Montroll, Frequency spectrum of crystalline solids, J.Chem.Phys., 10 (1942) 218
[Mor90] T. Morita, Cluster variation method and Möbius inversion formula, J.Stat.Phys., 59 (1990) 819
[Mul69] E.W. Müller and T.T. Tsong, Field Ion Microscopy, Principles and Applications, Elsevier, New York, 1969
[Nar89] W. Narkiewicz, Elementary and Analytic Theory of Algebraic Numbers, 2nd edition, Springer-Verlag, New York(1989)
[New56] J.R. Newman (ed.) The World of Mathematics, New York: Simon and Schuster, 1956
[Nin92] W. Ninham, B.D. Hughes, N.E. Frankel and M.L. Glasser, Möbius, Mellin, and Mathematical Physics, Physica, A186 (1992)441
[Ohn98] H. Ohnishi, Y. Kondo and K. Takayanagi, Quantized conductance through individual rows of suspended gold atoms, Nature 395(1998) 780
[Oma75] M.A. Omar, Elementary Solid State Physics, Addison-Wesley, Reading, MA, 1975
[Pan92] C.D. Pan and C.B. Pan, Elementary theory of numbers, Peking University Press, 1992
[Pan2005] C.D. Pan and C.B. Pan, Concise theory of numbers, Peking University Press,2005
[Pan2001] C.D. Pan and C.B. Pan, Theore of Algebraic Numbers, Shandong University Press, 2nd version, 2001
[Pla20] M. Planck, Nobel Lecture, 1920
[Pap95] T. Papas, The Music of Reason – Experiencing the Beauty of Mathematics through Quotations, Wide Publishing/Tetra, San Carlos, 1995
[Por2001] D.S. Portal, E. Artacho, J.Junquera, A. Carcia and J.M. Soler, Zigzag equibrium structure in monatomic wires, Surface Science, 482 (2001) 1261
[Qia2004] P. Qian, J. Shen and N.X. Chen, Phase stability and site preference of the rare-earth intermetallic compounds $R(Co,T)_{12}$ (R=Er, Dy; T=V, Ti, Cr, Mo, Mn, Nb, Ni, Cu), J.Alloy.Comp. 366 (2004) 41
[Rag87] H.A. Ragheb and M. Hamid, An approximation of Planck's formula for the inverse black body radiation, IEEE trans. AP-35(1987) 739
[Ram2000] J.P. Ramalho and G.V. Smirnov, On the structure of a local isotherm and solution to adsoption integral equation, Langmuir, 16 (2000) 1918
[Ree89] I.S. Reed, Y.Y. Choi, and X. Yu, Practical algorithm for computing the 2-D arithmetic Fourier transform, in Proc.SPIE Int.Soc.Opt.Eng., pp.54-61
[Ree90] I.S. Reed and D.W. Tufts, et al., Fourier Analysis and Signal Processing by the use of Möbius Inversion Formula, IEEE Trans. ASSP, 38 (1990) 458
[Ree92] I.S. Reed, M.T. Shih, T.K. Truong, E. Hendon and D.W. Tufts, A VLSI Architecture for Simplified Arithmetic Fourier Transform Algorithm, IEEE Trans. On Signal Processing, SP40 (1992) 1122

[Rem91] R. Remmert and R.B. Burckel, Theory of Complex Functions: Readings in Mathematics (1991), 125
[Ren91] S.Y. Ren and J.D. Dow, Generalized Mobius transforms for inverse problems, Phys. Lett., **A154** (1991) 215
[Ren93] G.B. Ren and N.X.Chen, Inversion of spontaneous magnetization for density of states of spin waves, JMMM 123(1993) L25
[Res85] R. Resnick and D. Halliday, Basic Concepts in Relativity and Early Quantum Theory, second ed., Wiley and sons, (1985) New York
[Ros93] H. Rosu, Mobius inverse problem for distorted black-holes, Nuovo Cimento Fisica,**B108** (1993) 1333
[Ros93a] H. Rosu, Black-holes and radiometry, Mod.Phys.Lett.A 8 (1993) 3429
[Ros84] J.M. Rose, J.R. Smith, F. Guinea, and J. Ferrante, Phys.Rev. B29 (1984) 2963
[Rot64] G.C. Rota, On the Foundations of Combinatorial Theory 1. Theory of Möbius Functions, Z. Wahrsch. Verw. Gebiete 2 (1964) 340-368
[Rus2001] V.A. Bushenkov, J.P. Ramalho, G.V. Smirnov, Adsoption Integral Equation via Complex Approximation with Constrains, J. Computational Chemistry, 22 (2001) 1058
[San78] J.M. Sanchez and D.de Fontaine, The fcc Ising model in cluster variation approximation, Phys.Rev.B17 (1978) 2926
[Sch61] S. Schweber, An Introduction to Relativistic Quantum Theory, Row-Peterson, New York, 1961
[Sch83] A.G. Schlijper, Convergence of cluster-variation method in the thermodynamic limit, Phys.Rev.B27(1983)6841
[Sch88] J.L. Shiff and R.W.Walker, A sampling theorem and Witner's results on Fourier coefficients, J.Math.Anal.Appl., 133 (1988) 466
[Sch92] J.L. Shiff, T.J.Surendonk and R.W.Walker, An algorithm for computing the inverse Z-transform,IEEE Trans. Signal Processing, SP 40 (1992) 2194
[Sch90] M.R. Schroeder, Number Theory in Science and Communication, 2nd ed, Springer-Verlag, New York (1990)
[She2003] Y.N. Shen, Z.D. Chen, X.H. Si, The general method to solve the inverse lattice problems in physics, J.Math. Anal. Appl., 279 (2003) 723
[She2004] J. Shen, P. Qian and N.X. Chen, Theoretical study on the structure for R2Co17(R=Y,Ce,Pr,Nd,Sm,Gd,Tb,Dy,Ho,Er) and R2Co17T(T=Be,C), J.Phys.Chen.Solid. 65 (2004) 1307
[Sip48] R. Sips, On structure of a catalyst surface, J.Chem.Phys.16 (1948) 490
[Sip50] R. Sips, On structure of a catalyst surface II., J.Chem.Phys.18 (1950) 1024
[Spe90] D. Spector, Supersymmetry and the Möbius-inversion function, Commun.Math.Phys., **127**(1990) 239
[Spe98] D. Spector, Duality, partial supersymmetry, and arithmetic number theory, J.Math.Phys., 39(1998) 1919
[Sti85] F.H. Stillinger and T.A.Weber, Computer simulation of local order in condensed phases of silicon, Phys.Rev.B. 31 (1985) 5262
[Sun87] X. Sun and D.L. Jaggard, The inverse blackbody radiation problem : A regularization solution, J.Appl.Phys., 62(1987) 4382

[Sun2008] V. Sundararaghavan and N.Zabaras, Weited multibody expansions for computing stable structures of multiatom systems, Phys.Rev. B77 (2008) 064101
[Sza47] O. Szasz, On Möbius inversion formula and closed sets of functions, Trans.Amer.Math.Soc., 62 (1947) 213
[Ter86] J. Tersoff, Reference levels for heterojunctions and Schottky barriers, Phys.Rev.Lett. 56 (1986) 632
[Tuf88] D.W. Tufts and G. Sadasiv, The Arithmetic Fourier Transform, IEEE ASSP Magazine, p.13 (January 1988)
[Tuf89] D.W. Tufts, Z. Fan and Z. Cao, Image processing and the arithmetic Fourier transform, SPIE High Speed Comput.II, Vol.1058, pp.46-53, 1989
[Vil97] F. Villieras, L.J. Michot, F. Bardot, J.M. Cases, M. Francois, W. Rudzinski, An improved derivative isothermo summation method to study surface hetrogeneity of cray minerals, Langmuir, 13 (1997) 1104
[Vzo92] M. Vzombathely, P. Brauer, M. Jaroniec, The solution of adsorption integral equation by means of the regularization, J. Comput. Chem., 13 (1992) 17
[Wan96] J.M. Wang and Y.Y. Zhou, Temperature Distributions of Accretion Disks in Active Galactic Nuclei, AstrophysJ, 469 (1996) 564
[Wan2005] C. Wang, Ab initio interionic potentials for CaO by multiple lattice inversion, J.Allows and Compounds, 388 (2005) 195
[Wan2010] Y.D. Wang and N.X. Chen, Atomistic study of misfit dislocation in metal/SiC(111) interfaces, J.Phys.: Condens. Matter. (2010) in print.
[Wei59] G. Weiss, On the inversion of the specific heat function, Prog.Theor.Phys.,**22** (1959) 526
[Wei82] M. Weinert, E. Wimmer and J. Freeman, Phys.Rev. B26 (1982) 4571
[Wei98] Y.C. Wei and N.X. Chen, Square wave analysis, J.Math.Phys.,39 (1998) 4226
[Wei2008] Y.G. Wei, X.M. You and H.F. Zhao, Investigations of cohesive zone properties of the interface between metal and ceramic substrate at nano and micro-scales heterogeneous material mechanics, pp.141-144, 2008
[Win47] A. Wintner, An Arithmetical Approach to Ordinary Fourier Series, Botimore, 1947
[Wu2003] C.S. Wu, The mathematical physics, Peking University Press,2003
[Xie91] T.L. Xie, P.F. Goldsmith and W. Zhou, A new method for analyzing IRAS data to determine the dust temperature distribution, AstrophysJ., **371** (1991) L81
[Xie93] T.L. Xie, P.F. Goldsmith, R.L. Shell and W. Zhou, Dust temperature distributions in star-forming condensations, AstrophysJ., **402** (1993) 216
[Xie94-1] Q. Xie and M.C. Huang, Application of lattice inversion method to embedded-atom method, Phys.Ptat.Sol., B186 (1994) 393
[Xie94-2] Q. Xie and M.C. Huang, A lattice inversion method to construct the alloy pair potential for the embedded-atom method, J.Phys: Condens. Matter. 6 (1994) 11015
[Xie95a] Q. Xie and N.X. Chen, Unified inversion technique for fermion and boson integral equations, Phys.Rev.,E52(1995) 351

[Xie95b] Q. Xie and N.X. Chen, Matrix-inversion: Applications to Möbius inversion and deconvolution, Phys.Rev.E52 (1995) 6055
[Xie95] Q. Xie and N.X. Chen, Recovery of an N-body potential from a universal cohesion equation, Phys.Rev. B51 (1995) 15856
[Xu94] T.F. Xu, On the uniqueness of the inverse blackbody radiation problem, Phys.Lett. A196 (1994) 20
[You80] R.M. Yang, An Introduction to Nonharmonic Fourier Series, Academic Press, 1980 New York
[Yan98] A.I. Yanson, G.R. Bollinger, H.E. vandenBrom and N. Agra, Formation and manipulation of a metallic wire of single gold atoms, Nature, 395 (1998) 783
[Zha98] W.Q. Zhang, Q. Xie, X.D. Zhao and N.X. Chen, Lattice-inversion embedded-atom model and its applications, Science in China A28 (1998) 183 (in Chinese)
[Zha2002] S. Zhang and N.X. Chen, *Ab initio* interionic potentials for NaCl by multiple lattice inversion, Phys.Rev. B66 (2002) 064106
[Zha2003a] S. Zhang and N.X. Chen, Lattice inversion for interionic potentials, J.Chem.Phys. 118(2003) 3974
[Zha2003b] S. Zhang and N.X. Chen, Determination of the B1-B2 transition path RbCl by Mobius pair potentials, Philos.Mag. 83 (2003) 1451
[Zha2003c] S. Zhang and N.X. Chen, Energies and stabilities sodium chloride clusters based on inversion pair potentials, Physica B325 (2003) 172
[Zha2005] S. Zhang and N.X. Chen, Lattice inversion for interionic potentials in AlN, GaN and InN, Chem.Phys., 309 (2005) 309
[Zha2008] H.Y. Zhao and N.X. Chen, Inverse adhesion problem for extracting interfacial pair potentials for Al(001)/3C-SiC(001) interface, Inverse Problems, 24 (2008) 035019
[Zha2009] H.Y. Zhao, N.X. Chen, Y.Long, Interfacial potentials for Al/SiC(111), J. Phys: Condens.Matter. 21 (2009) 225002
[Zha2009a] L. Zhang and C.B. Zhang and Y. Qi, Molecular dynamics study on structural change of a Au-959 cluster supported on MgO(100) surface at low temperature, Acta Physics Sinica, 58 (2009) S53

Index

$Dirac - \delta$ function, 95
2D AFT algorithm, 164
2D square lattice, 152
3D inverse lattice problem, 187

A.M.Schröflies (1853–1928), 183
Abel, 69
additive biorthogonality, 104
additive Cesàro inversion formula, 122
alternative Möbius series inversion, 21
Arithmetic Fourier Transform, 126
arithmetic function, 8
arithmetic functions of second kind, 79

Bazant and Kaxiras, 186
Bessel derivatives, 110
biothogonality, 96
body-centered cubic lattice, 192
Bohr magneton, 73
Bojarski, 50
Bose, 53
Bruns' algorithm, 126

Carlsson, Gellat and Ehrenreich (CGE), 184
Cesàro-Möbius inversion formula, 36
Chebyshev, 29
Chengbiao Pan or Pan Chengbiao, xi
closeness of multiplicative operations, 187

Cluster expansion method, 245
congruence analysis, 149
Cooley and Tukey, 126
coordination number, 161

Debye approximation, 60
Dirichlet inverse, 11
Dirichlet product, 9
divisor function $\tau(n)$, 13
dual orthogonality, 100

E.S.Fedorov (1853–1919), 183
Einstein approximation, 60
Eisenstein integer, 173
embedded-atom method (EAM), 211
Euler $\varphi-$ function, 16

face-centered cubic structure, 189
Fermi integral equation, 86
Feynman, 252
field-ion microscopy (FIM), 197
Fourier deconvolution, 83

Gödel numbering, 33
Gaussian integers, 154
graphene, 149
Guo Dunren or Dunren Guo, xi

Interatomic potentials between atoms across interface, 215
Interval and chain, 235
inverse heat capacity problem, 60

inverse interfacial adhesion problem, 215
inverse transmissivity problem, 75

Jeans, 50

L1$_2$ structure, 195
Langmuir, 89
Laplace operator, 74
Legendre derivatives, 109
Local finite POSET, 236

Möbius function, 7
Möbius function and Witten index, 32
Möbius function of Gaussian integers, 159
Möbius function on locally finite POSET, 237, 239
Möbius inverse formula, 17
Möbius inverse formula on Eisenstein integers, 178
Möbius inverse formula on Gaussian integers, 160
Möbius series inversion formula, 7
Möbius series inversion formula of second kind, 82
Möbius, Chebyshev and modulation transfer function, 28
Möbius–Chebyshev inverse formulas, 29
Möbius-Cesàro inverse formula, 37
Möbius-Cesàro inversion formula of the second kind, 119
Max von Laue, 183
Menghui Liu or Liu Menghui, xi
modulation transfer function, 29
Montroll, 55
Montroll and Lifshitz, 54
multiplicative function, 12

nanowires, 149

P.P. Weald, 183
partially ordered set (POSET), 234
Planck, 50
power function $\iota_s(n)$, 13
principle of inclusion and exclusion, 241

quadrant, 158

Ramanujun sum, 140
Rayleigh, 50
Reed, 126, 132
relaxation-time distribution, 90
Riemann's zeta function, 25
Rota, 252

stretching curve, 210
sum rule, 17

temperature dependence of Debye temperature, 62
temperature dependence of Einstein temperature, 67
Tikhonov, 67
totally ordered set (TOSET), 233
Tufts and Sadasiv, 126

uniformly sampling, 146
unique factorization, 4
unique factorization theorem, 155

VLSI architecture, 172

W.H. Bragg and W.L. Bragg, 183
Walsh function, 102
Wang Yuan or Yuan Wang, xi
Weiss, 62
Wien, 50
Wiener–Hopf method, 89
Wintner, 126
Witten, 32